Advances in Robust Control and Applications

Edited by Mahmut Reyhanoglu

Published in London, United Kingdom

Advances in Robust Control and Applications
http://dx.doi.org/10.5772/intechopen.1000358
Edited by Mahmut Reyhanoglu

Contributors
Abdallah Laidi, Abdul Kareem, Aissa Benhammou, Akshatha Naik, Aliasghar Arab, Anton de Ruiter, Farana Mohammed Imran, Mahmut Reyhanoglu, Mohamed Amine Hartani, Mohammed Tajudeen Jimoh, Sean Kalaycioglu, Sookshma Adiga, Varuna Kumara

First published in London, United Kingdom, 2025 by IntechOpen
IntechOpen is the global imprint of INTECHOPEN LIMITED, registered in England and Wales, registration number: 11086078, 167-169 Great Portland Street, London, W1W 5PF, United Kingdom

For EU product safety concerns: IN TECH d.o.o., Prolaz Marije Krucifikse Kozulić 3, 51000 Rijeka, Croatia, info@intechopen.com or visit our website at intechopen.com.

British Library Cataloguing-in-Publication Data
A catalogue record for this book is available from the British Library

Advances in Robust Control and Applications
Edited by Mahmut Reyhanoglu
p. cm.
Print ISBN 978-1-83634-068-3
Online ISBN 978-1-83634-067-6
eBook (PDF) ISBN 978-1-83634-069-0

Meet the editor

Mahmut Reyhanoglu is currently the Director of Robotics Engineering at Columbus State University (CSU) in Georgia, USA. Prior to joining the CSU, he was the Glaxo Wellcome Distinguished Professor of Engineering at the University of North Carolina at Asheville in North Carolina, USA. His extensive research uses advanced mathematical techniques and models arising from fundamental physical principles. His primary research interests are in the areas of nonlinear dynamical systems and control theory, with particular emphasis on applications to mechanical and aerospace systems, robotics and mechatronics. He has edited 5 books and authored/co-authored 5 book chapters and over 140 peer-reviewed journal/proceedings papers. He served as an Associate Editor on the IEEE Transactions on Automatic Control Editorial Board and the IEEE Control Systems Society Conference Editorial Board. He also served as an International Program Committee member for several conferences and as a member of the AIAA Guidance, Navigation, and Control Technical Committee. He is currently an editor of the *International Journal of Aerospace Engineering and the MDPI Electronics Journal.*

Contents

Preface

Recently, there has been considerable excitement over the emergence of new mathematical techniques for modeling and analyzing complex dynamic systems. Such complex systems are prone to unwanted effects of uncertainties, disturbances, and noise. The growing availability of more powerful and affordable microprocessors, coupled with the demand for improved performance, encouraged control engineers to develop innovative and robust nonlinear control algorithms to address the challenges related to uncertainties, disturbances, and noise. This book reviews recent advancements in robust control and applications.

Chapter 1 is the introductory chapter, briefly describing recent advances in robust control approaches and their applications in various engineering fields.

Chapter 2 presents a general introduction to Model Predictive Control (MPC), focusing on linear, unconstrained state space MPC. The linear quadratic objective function forms the basis of MPC, followed by a discussion of its optimal solution. The formulation of the prediction equation for state space MPC is outlined, with a detailed examination of the terms involved in defining the optimal control law. Additionally, the design of two types of observers for MPC is explored: the predictor-only type and the predictor-corrector type.

Chapter 3 is devoted to modeling robotic vehicles, emphasizing a comprehensive and detailed description of their complex dynamics. It considers various challenges autonomous vehicles encounter, including nonlinearities, side slips, and unpredictable environmental disturbances, to offer a more realistic framework for understanding vehicle behavior. Central to this method is the development of a robust Nonlinear Model Predictive Controller (NMPC) specifically designed to address these issues. By utilizing advanced control techniques, the NMPC effectively manages the vehicle's stability, ensuring smooth and precise operation, even in the face of uncertainties like varying road friction, sensor noise, and external forces such as wind or uneven terrain.

Chapter 4 combines Nonlinear Model Predictive Control (NMPC), Nonlinear AutoRegressive eXogeneous inputs (NARX) based learning, and passivity constraints to create a highly adaptive, robust, and scalable control framework for space systems. Important contributions include the application of NARX neural networks for adaptive state estimation and passivity-based NMPC for robust control. These methods pave the way for enhanced autonomy, stability, and efficiency in space operations, establishing a critical foundation for the future of intelligent spacecraft and robotic missions.

Chapter 5 deals with the Sliding Mode Control (SMC) technique. It presents a moving, sliding surface whose slope is real-time adjusted using a fuzzy logic controller. This adjustment allows the sliding surface to be rotated in a manner that enhances the dynamic performance of the control system. Computer simulations for electronic throttle control demonstrate that the proposed control scheme achieves better dynamic performance than the conventional super-twisting sliding mode controller with a fixed sliding surface.

Finally, Chapter 6 presents simulation results for standalone energy systems controlled by Proportional-Integral and Super Twisting SMC techniques. The effectiveness of these two techniques is compared by using Mean Absolute Error (MAE) and Root Mean Square Error (RMSE) performance metrics.

Mahmut Reyhanoglu, Ph.D.
Robotics Engineering Laboratory,
Columbus State University,
Columbus, Georgia, USA

Chapter 1

Introductory Chapter: Advances in Robust Control and Applications

Mahmut Reyhanoglu

1. Introduction

There has been considerable excitement recently over the emergence of new mathematical techniques for modeling and analyzing complex dynamic systems. Such complex systems are prone to unwanted effects of uncertainties, disturbances, and noise. The growing availability of more powerful and affordable microprocessors, coupled with the demand for improved performance, encouraged control engineers to develop innovative and robust nonlinear control algorithms to address challenges related to uncertainties, disturbances, and noise. This introductory chapter briefly describes the advances in robust control approaches and their applications in various engineering fields.

Sliding mode control (SMC) is a widely adopted nonlinear robust control technique. SMC relies on high controller gains to ensure robust performance in the presence of disturbances and uncertainties. However, the traditional SMC technique suffers from unwanted chattering effects in the system response [1]. Recently introduced modified versions of SMC include the learning-based SMC [2] and the neuro-adaptive SMC approaches [3].

An alternative method to handle disturbances and uncertainties is the learning-based model predictive control approaches [4]. Neural network (NN)-based learning is a popular control technique with different variants such as NN-based online learning [5] and NN-based adaptive control [6, 7]. Another approach amalgamates a fuzzy neural network with a conventional Proportional plus Derivative controller (FNN + PD) [8, 9].

Another widely used approach is the feedback linearization control (FLC), though its performance diminishes under uncertainties and disturbances, leading to a failure in convergence to zero for closed-loop errors [10]. To address this, FLC with integral action has been proposed to guarantee steady-state error convergence. Several other learning-based approaches have been developed in conjunction with FLC to overcome its limitations [11–13]. In [13], an artificial NN with a nonlinear auto-regressive-moving average (ARMA) model is designed to enable the learning of nonlinear system inputs. A dynamic linearization approach has been introduced for discrete-time nonlinear systems [14]. A gradient-descent-based learning gain and disturbance observer for nonlinear systems is introduced in [15]. While these learning-based nonlinear methods offer several advantages, such as handling uncertainties, unmodeled dynamics, and disturbances, their real-time implementation remains difficult due to the high computational demands. To overcome this challenge, [16] extends the fundamental steps from previous works [17–19] to create a simple learning (SL)-based control strategy for trajectory tracking in a specific class of dynamical systems.

IntechOpen

The SL-based method is used to design the traditional feedforward control law based on the system's nominal model. The strategy also uses the gradient-descent method to minimize the desired closed-loop error function and derive adaptation rules for disturbance estimates and control gains.

2. Applications of robust control techniques

There are numerous applications of sliding mode control (SMC) schemes, including control of chemical processes, electric motors, automatic flight control, control of space systems and autonomous robots [1]. Application of learning-based SMC and the neuro-adaptive SMC approaches to robotic systems can be found in [2, 3].

The learning-based model predictive control approaches have been successfully applied to autonomous vehicles, robot manipulation, multi-robot systems, energy systems, chemical and industrial processes, flight controls, agriculture and environmental systems [4]. Neural network (NN)-based learning techniques utilize neural networks' capabilities to model complex systems while optimizing control decisions in real time. NN-based online learning approaches are based on continuous parameter updates as new data becomes available. Important applications of NN-based online learning include adaptive path planning for autonomous driving, autonomous flight of unmanned aerial vehicles (UAVs), robotic process automation, optimal control of smart grids, robotic surgeries, and economic system modeling [5]. NN-based adaptive control techniques enable control systems to adapt to time-varying system dynamics, uncertainties, and disturbances. Some of the applications NN-based adaptive control are process optimization, smart grids, signal processing, cybersecurity, control of robot manipulators, autonomous vehicles, and UAVs [6, 7]. Fuzzy Neural Network plus Proportional Derivative (FNN + PD) control employs SMC-based learning technique for robust control. The FNN + PD control has proven a powerful technique for the control of shipborne manipulators [8] and fixed-wing UAVs [9].

The learning-based feedback linearization control (FLC) technique, which combines machine learning with feedback linearization, has been widely applied in various important engineering fields, ranging from robotics to communication and network systems [10–15]. Recently, the simple learning (SL) based control strategy has been applied to a general class of dynamical systems [16]. In particular, the SL-based control approach is applied to the control of several important engineering systems including electric pumps [17], UAVs [18], and 2-DOF helicopter [19]. **Figure 1** shows the flow diagram of SL-based control strategy applied to a tricopter

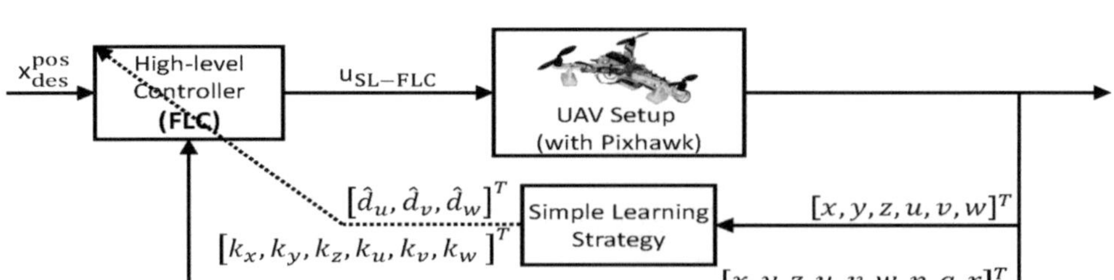

Figure 1.
The flow diagram of SL-based control strategy [18].

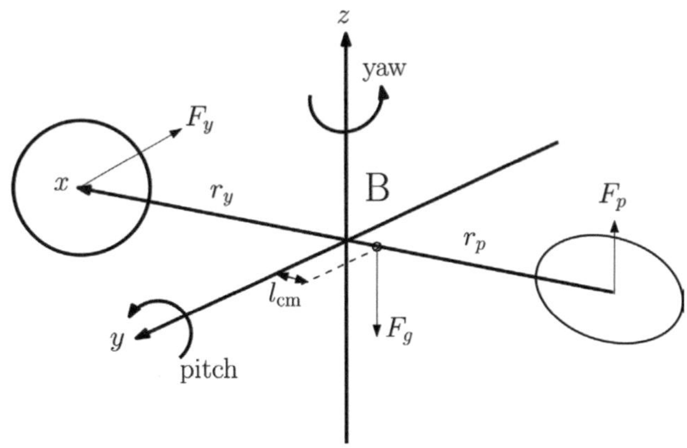

Figure 2.
Schematic representation of a 2-DOF helicopter [19].

UAV. Schematic representation of the 2-DOF helicopter is shown in **Figure 2**. Finally, in [20], a machine learning-based approach is implemented to address challenges caused by disturbances and uncertainties arising in the control of autonomous solar panel cleaning systems.

Author details

Mahmut Reyhanoglu
Columbus State University, Columbus, Georgia, USA

*Address all correspondence to: mreyhanoglu@icloud.com

References

[1] Edwards C, Spurgeon S. Sliding Mode Control. Theory and Applications. London: Taylor and Francis; 1998

[2] Kayacan E. Sliding mode learning control of uncertain nonlinear systems with Lyapunov stability analysis. Transactions of the Institute of Measurement and Control. 2019;**41**:1750-1760

[3] Topalov AV, Kaynak O. Online learning in adaptive neurocontrol schemes with a sliding mode algorithm. IEEE Transactions on Systems, Man, and Cybernetics, Part B Cybernetics. 2001;**31**:445-450

[4] Kayacan E, Kayacan E, Ramon H, Saeys W. Learning in centralized nonlinear model predictive control: Application to an autonomous tractor-trailer system. IEEE Transactions on Control Systems Technology. 2015;**23**:197-205

[5] Kocer BB, Tjahjowidodo T, Seet GGL. Centralized predictive ceiling interaction control of quadrotor VTOL UAV. Aerospace Science and Technology. 2018;**76**:455-465

[6] Fu C, Hong W, Zhang L, Guo X, Tian Y. Adaptive robust backstepping attitude control for a multi-rotor unmanned aerial vehicle with time-varying output constraints. Aerospace Science and Technology. 2018;**78**:593-603

[7] Wu B, Wu J, He W, Tang G, Zhao Z. Adaptive neural control for an uncertain 2-DOF helicopter system with unknown control direction and actuator faults. Mathematics. 2022;**10**:4342

[8] Xu Z, Li W, Wang Y. Robust learning control for shipborne manipulator with fuzzy neural network. Frontiers in Neurorobotics. 2019;**13**:11

[9] Kayacan E, Khanesar MA, Rubio-Hervas J, Reyhanoglu M. Learning control of fixed-wing unmanned aerial vehicles using fuzzy neural networks. International Journal of Aerospace Engineering. 2017;**2017**:5402809

[10] Kayacan E, Fossen TI. Feedback linearization control for systems with mismatched uncertainties via disturbance observers. Asian Journal of Control. 2019;**21**:1064-1076

[11] Jafari M, Xu H, Garcia Carrillo LR. A neurobiologically-inspired intelligent trajectory tracking control for unmanned aircraft systems with uncertain system dynamics and disturbance. Transactions of the Institute of Measurement and Control. 2019;**41**:417-432

[12] Jafari M, Xu H. Intelligent control for unmanned aerial systems with system uncertainties and disturbances using artificial neural network. Drones. 2018;**2**:30

[13] Sahin S. Learning feedback linearization using artificial neural networks. Neural Processing Letters. 2016;**44**:625-637

[14] Chi R, Hou Z, Huang B, Jin S. A unified data-driven design framework of optimality-based generalized iterative learning control. Computers and Chemical Engineering. 2015;**77**:10-23

[15] You S, Son YS, Gui Y, Kim W. Gradient-descent-based learning gain for backstepping controller and disturbance observer of nonlinear systems. IEEE Access. 2023;**11**:2743-2753

[16] Reyhanoglu M, Jafari M. A simple learning approach for robust tracking control of a class of dynamical systems. Electronics. 2023;**12**:2026

[17] Jafari M, Reyhanoglu M, Kozhabek Z. Simple learning-based robust nonlinear control of an electric pump for liquid-propellant rocket engines. Electronics. 2023;**12**:3527

[18] Mehndiratta M, Kayacan E, Reyhanoglu M, Kayacan E. Robust tracking control of aerial robots via a simple learning strategy-based feedback linearization. IEEE Access. 2019;**8**:1653-1669

[19] Reyhanoglu M, Jafari M, Rehan M. Simple learning-based robust trajectory tracking control of a 2-DOF helicopter system. Electronics. 2022;**11**:2075

[20] Hajiahmadi F, Jafari M, Reyhanoglu M. Machine learning-based control of autonomous vehicles for solar panel cleaning systems in agricultural solar farms. Agricultural Engineering. 2024;**6**:1417-1435

Chapter 2

Perspective Chapter: Basic Formulation of Unconstrained Linear State-Space Model Predictive Control

Mohammed Tajudeen Jimoh

Abstract

Model predictive control (MPC) is a receding horizon type of control system that uses the internal model in its control action. Different types of model formats that may be linear or nonlinear are employed in its design. Its design involves systematic formulation of the model for output prediction and optimized input calculations, even for multivariable and/or constrained systems. This makes MPC varied and complex in its formulation and implementation. In this chapter, the focus is on linear, unconstrained state-space MPC, which is a good starting point for beginners to MPC. The chapter provides details of how, using the governing objective function, the optimal control law is formulated. More importantly, it provides details of how the terms involved in the resulting optimal control law are formulated and evaluated. In addition, the formulation of two types of observers for the state-space MPC is presented. By using a benchmark model and a MATLAB code, a simplified step-by-step implementation guide of the MPC is provided.

Keywords: model predictive control, constrained MPC, quadratic objective function, linear MPC, Kalman filter, observer

1. Introduction

Model predictive control (MPC) is an advanced control system that relies on the use of an internal dynamic model of the plant being controlled to achieve its control action. The origin of MPC and its early application have been traced to the 1970s in the process industries, particularly oil refining and petrochemicals [1, 2]. The slow dynamics of the process industries were considered just right for the heavy computational requirements of MPC. Many other industries (food processing, pulp and paper, mining and metallurgy, defense, and automotive) have since embraced the technology [3–5]. Even in robotics and aerospace industries, where the dynamics are very fast and nonlinear, with the diversity of mission platforms, MPC has found wide applications [6–8]. This has been made possible by the advances in memory and computational power of computers.

MPC is particularly popular because of the ease with which it can be applied to multivariable systems, and its ability to handle constraints almost effortlessly. It can also account for actuator limitations and can be used for systems with complex dynamics with long delay times or of non-minimum phase unstable processes. Another attraction is that it is based on an open methodology underlined by principles that allow for future extensions [2]. In general, MPC control action is highly efficient and is able to operate for long periods of time with hardly any intervention.

It, however, has some disadvantages. The formulation of its control law is more complex compared to classical PID, and it requires a high computation load [2]. It can be operationally difficult, and it requires high installation and maintenance costs. It lacks flexibility as it is plant-specific [9].

1.1 General principle of model predictive control

The MPC control action is carried out using three basic elements, as illustrated in **Figure 1**. The MPC uses an internal model (linear or nonlinear), together with past outputs/inputs/states, to construct a future output prediction equation $\hat{Y}$ for a prediction horizon. The predicted future output is combined with set point S (or reference T) trajectory information, and possibly information about input and output constraints, to formulate the necessary objective (or cost) function J. Then the cost function is solved to obtain a control law from which a set of optimum control moves $\Delta \breve{U}$ for a control horizon is calculated. If constraints are not accounted for in the formulation of the objective function, the objective function is linear and can be solved explicitly using the linear least squares methods. For a cost function that includes constraints, the optimal control moves are obtained using quadratic programming (QP) techniques [10].

The control moves are calculated at a regular time interval Δt. They are calculated such that the predicted future outputs $\hat{Y}$ optimally track defined (set point) reference trajectory T within a prescribed horizon P. The control moves are calculated based on the receding horizon strategy. This receding horizon strategy is defined as this: at a

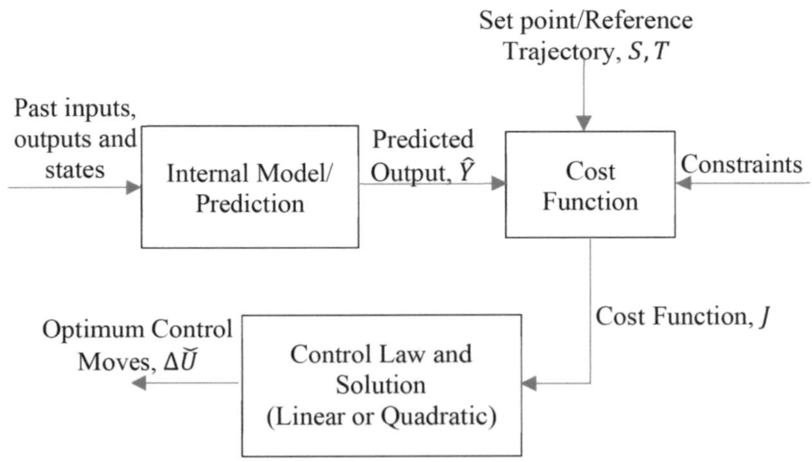

Figure 1.
Elements of a basic MPC.

sampling instant k, a sequence of future H optimal control moves $\breve{U}$ is calculated. From this sequence, only the first control move $\breve{u}\ (k)$ is sent to the plant as the manipulated variable $u(k)$, and the others are neglected. Then, the same procedure is repeated at the next sampling instant $k+1$. In MPC, P is referred to as the prediction horizon while H is the control horizon. Both are important MPC tuning parameters. For constrained systems, a large value of P allows the MPC to plan well ahead and avoid or minimize the effect of the constraints. Likewise, for systems with plant delay, where t_d is the equivalent sampling instant of the time delay, a large value of P $(P > > t_d)$ allows the MPC to accommodate the delay in its predictions. For a stable system, P is recommended to be equal to or greater than the open loop settling time of the process being controlled [11]. On the other hand, for a system with a time delay, the value of H is usually set such that $H < < P - t_d$. This allows a unique set of optimal control moves to be determined at every sampling instant [12, 13].

1.2 Major types of MPC

The major available MPC types can be broadly categorized into three: the Finite Impulse/Step Response MPC, the Transfer function MPC, and the state-space MPC. The MPC type is actually determined by the format of its internal model, which in turn determines the nature of the prediction equation that is formulated.

The Finite Impulse/Step Response MPC, which is generally referred to as Dynamic Matric Control (DMC), is the first of the MPC types to be reported [14]. This type uses linear discrete-time empirical finite impulse/step response models in the formulation of its prediction equation. It is limited to stable processes and processes without integrators. Discussions on the development of prediction equations for step response MPC are widely available. See, for example, Refs. [15–17].

Transfer function MPC, broadly referred to as Generalized Predictive Control (GPC), was first reported in the 1980s [18]. Its prediction equation is based on the transfer function (or input-output difference equation) model. It is suitable for stable and unstable processes. Details about the formulation of the prediction equation for this type of MPC can be found in Refs. [2, 10, 19].

State-space MPC was introduced in the late 1980s. The first account and description of this form of MPC, according to Qin and Badgwell [1], were contained in Refs. [20, 21]. As the name implies, state-space MPC uses discrete-time state-space models for the derivation of its prediction equation.

State-space has become the most common approach to MPC design [22]. The state-space internal model can accommodate different problem dynamics (integrating, stable and unstable) easily and it can handle multivariable processes most effortlessly [10]. In addition, a large collection of modern control theory and analysis methods can be applied to the study of its algorithm, robustness, and stability.

State-space MPC is, however, complicated by the need to incorporate a state estimator, commonly referred to as an observer, in its prediction [16]. This is because, in implementation, the internal model used is not always a perfect representation of the dynamics of the process being controlled. That is there is always some degree of model-plant mismatch. Also, not all the states of the process upon which predictions are made are directly measurable. Yet good information on the states of the process is required for accurate prediction by state-space MPC. The observer is normally included to ensure good control actions.

Despite its widespread use, state-space MPC has attracted some criticisms: One obvious criticism is that certain processes may require very large state matrices to fully describe them and that this may lead to numerical problems during computations. Another criticism is that a state-space model is not very transparent: It is most times difficult to see what the figures are saying, except for very experienced engineers. However, some researchers are now showing that some of these criticisms are not completely justified. For example, Maciejowski [10] showed that a complicated multivariable step response model with input delays can be converted to an equivalent state-space model. There are also works on state reduction. A simple algorithm for obtaining state-space dynamic from step response data is given in Ref. [23].

State-space MPC formulations are well documented. See for instance Refs. [2, 10, 11, 24–29] and many others.

1.3 Chapter objective

The major hindrances to wider application of MPC generally are its opaqueness and complex design process. Their implementation on multivariable process adds to the complexity. These reasons have limited its study even for research purposes. Many trainings and researches on the MPC rely mainly on proprietary MPC software. The focus of this chapter is on presenting a simplified formulation of unconstrained, state-space MPC, with emphasis on its key aspects: the prediction equation, the objective function, the control law, and the observer. The formulations are based on the generalized Multiple-Input-Multiple-Output (MIMO) system. In the formulations, the variables and parameters involved, as well as the dimensions of the associated matrices and vectors, are defined.

2. Linear, unconstrained state-space MPC and its formulation

Linear state-space MPC uses linear, time-invariant dynamic state-space models in the design of its objective function [25]. An unconstrained MPC does not include provisions for handling constraints on the inputs, the inputs changes, and the outputs in its formulation.

The basic flow of signal in a state-space MPC is shown in **Figure 2**. The MPC uses set point information s, together with an estimated state of the plant $\hat{x}$, to calculate the control move u, that is sent to the plant. The observer makes use of the measured plant output y, and the previous plant input to predict the next state of the plant.

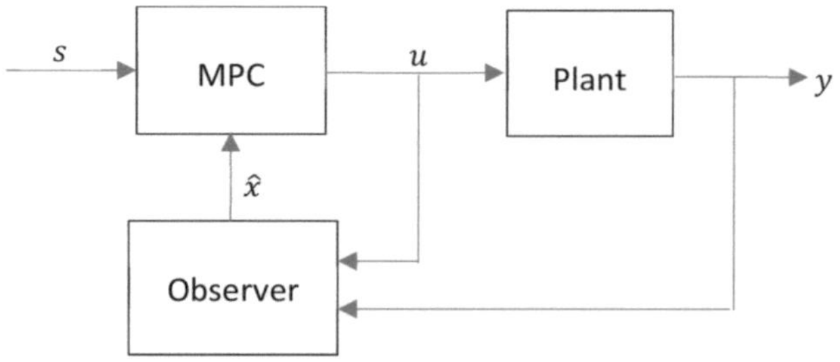

Figure 2.
Basic signal flow of a plant controlled with state-space MPC.

2.1 The output prediction equation

For a multi-input, multi-output (MIMO) system of m outputs and w inputs, its dynamics may be represented in proper (a proper system is one in the current input has no effect on the output), discrete-time, state-space equation as: [11, 30, 31]:

$$x(k+1) = A_d x(k) + B_d u(k) + G_d z(k) \tag{1}$$

$$y(k) = C_d x(k) + v(k) \tag{2}$$

If the system has n states, the system matrix A_d, the input matrix B_d, and the output (measurement) gain matrix C_d have dimensions $A_d \in \mathbb{R}^{n\times n}$, $B_d \in \mathbb{R}^{n\times w}$, $C_d \in \mathbb{R}^{m\times n}$, respectively. Also, the predicted state vector x, the predicted output vector $\hat{y}$, and the input vector u have dimensions $x \in \mathbb{R}^{n\times 1}$, $y \in \mathbb{R}^{m\times 1}$, and $u \in \mathbb{R}^{w\times 1}$, respectively. $z = \in \mathbb{R}^{n\times 1}$ is the process noise vector, $v = \in \mathbb{R}^{m\times 1}$ is the measurement noise vector, and $G_d = \in \mathbb{R}^{n\times n}$ is a diagonal (usually identity) matrix relating the process noise to the state variables.

The equivalent augmented model of the discrete-time, state-space system that is used in state-space MPC formulation, is given as [11]:

$$\begin{bmatrix} \Delta x(k+1) \\ y(k+1) \end{bmatrix} = \begin{bmatrix} A_d & 0_{n,m} \\ C_d A_d & I_m \end{bmatrix} \begin{bmatrix} \Delta x(k) \\ y(k) \end{bmatrix} + \begin{bmatrix} B_d \\ C_d B_d \end{bmatrix} \Delta u(k) \tag{3}$$

$$y(k) = [0_{m,n} \;\; I_m] X(k) \tag{4}$$

or simply as:

$$X(k+1) = AX(k) + B\Delta u(k) \tag{5}$$

$$y(k) = CX(k) \tag{6}$$

where $A, B,$ and C are the augmented matrices with the following dimensions: $A \in \mathbb{R}^{(n+m)\times(n+m)}$, $B \in \mathbb{R}^{(n+m)\times w}$, $C \in \mathbb{R}^{m\times(n+m)}$. The new augmented state matrix X has dimension $X \in \mathbb{R}^{(n+m)\times 1}$ and the input change vector $\boldsymbol{\Delta u}$ has the dimension $\boldsymbol{\Delta u} \in \mathbb{R}^{w\times 1}$. The matrix $0_{n,m}$ is an all-zeros elements matrix of n rows and m columns, while $0_{m,n}$ is the transpose of $0_{n,m}$. The matrix I_m is an identity matrix of dimension m.

Using the augmented state models of Eqs. (3)–(6) the output prediction equation within the prediction horizon P and control horizon H for state-space MPC can be written in generalized vector-matrix form as [11]:

$$\underbrace{\begin{bmatrix} \hat{y}(k+1) \\ \hat{y}(k+2) \\ \hat{y}(k+3) \\ \vdots \\ \hat{y}(k+P) \end{bmatrix}}_{\hat{Y}} = \underbrace{\begin{bmatrix} CB & 0 & \cdots & 0 \\ CAB & CB & \cdots & 0 \\ CA^2B & CAB & \cdots & 0 \\ \vdots & \vdots & \ddots & \vdots \\ CA^{P-1}B & CA^{P-2}B & \cdots & CA^{P-M}B \end{bmatrix}}_{\Phi} \underbrace{\begin{bmatrix} \Delta u(k) \\ \Delta u(k+1) \\ \Delta u(k+2) \\ \vdots \\ \Delta u(k+H-1) \end{bmatrix}}_{\Delta U} + \underbrace{\begin{bmatrix} CA \\ CA^2 \\ CA^3 \\ \vdots \\ CA^P \end{bmatrix}}_{\Gamma} X \tag{7}$$

or simply as:

$$\hat{Y} = \Phi \Delta U + \Gamma X \tag{8}$$

where $\Phi \in \mathbb{R}^{mP\times wH}$ and $\Gamma \in \mathbb{R}^{mP\times(n+m)}$ can be calculated as constant matrices from the augmented state-space matrices.

The vector $\hat{Y} \in \mathbb{R}^{mP\times 1}$ is the vector of predicted outputs over the prediction horizon P. Each of $\hat{y}(k+i)$ in $\hat{Y}$ is a vector defined as:

$$\hat{y}(k+i) = \left[\hat{y}_1(k+i), \hat{y}_2(k+i), \dots, \hat{y}_m(k+i)\right]^T \tag{9}$$

The vector $\boldsymbol{\Delta U} \in \mathbb{R}^{wH\times 1}$ is also a vertical concatenation of vectors Δu_{k+j-1} over the control horizon. Each of $\boldsymbol{\Delta u}(k+j-1)$ $(j = 1, 2, \dots, H)$ in $\boldsymbol{\Delta U}$ is a vector defined as:

$$\boldsymbol{\Delta u}(k+j-1) = [\boldsymbol{\Delta u}_1(k+j-1), \boldsymbol{\Delta u}_2(k+j-1), \dots, \boldsymbol{\Delta u}_w(k+j-1)]^T \tag{10}$$

2.2 MPC objective function

The popular quadratic objective function used in the formulation of the optimal control law for most types of MPC is given as [2]:

$$\min_{\Delta u} J = \sum_{i=1}^{P} q_i \left[y_i(k+i) - t_i(k+i)\right]^2 + \sum_{j=1}^{H} r_j \left[(\Delta u_j(k+j-1)\right]^2 \tag{11}$$

where y_i is the measured output, t_i is the reference signal, and Δu_j is the input signal. $(q_i \geq 0)$ is the output penalty weight constant, and $(r_j > 0)$ is an input penalty weight constant. The weights are used to either increase or decrease considerations given to the variables $\left[y_i(k+i) - t_i(k+i)\right]$ and Δu_j during optimization.

In MPC, a high q_i usually results in faster tracking of the output. On the other hand, a high value of r_j on an input helps to suppress its aggressive behavior. The reference trajectory t_i is an approximation used to achieve a gradual transition from the current measurement y_i to the set point s_i during optimization.

For a generalized linear MIMO system with m outputs and w inputs, the cost function can be written in matrix form as [2, 32]:

$$\min_{\Delta \breve{U}} J = \left(T - \hat{Y}\right)^T \mathrm{Q}\left(T - \hat{Y}\right) + (\Delta U)^T \mathrm{R}(\Delta U) \tag{12}$$

where $T \in \mathbb{R}^{mP\times 1}$ is a vector of defined reference trajectory within the prediction horizon, and $\hat{Y} \in \mathbb{R}^{mP\times 1}$ is a vector of predicted outputs within the prediction horizon (Eqs. (7) and (8)). $\Delta U \in \mathbb{R}^{wH\times 1}$ is a vector of the input change within the control horizon. $\mathrm{Q} \in \mathbb{R}^{mP\times mP}$ and $\mathrm{R} \in \mathbb{R}^{wH\times wH}$ are square penalty matrices whose elements are derived from the weights on the outputs and input channels, respectively.

2.2.1 Output and input penalty weight matrices

If the weights on the m output channels are specified as a vector $q = \left[q_1, q_2, \dots, q_\mathrm{m}\right]$, and the weights on the w input channels are given as a vector $r = \left[r_1, r_2, \dots, r_\mathrm{w}\right]$, then the constant penalty matrix Q and R are defined, respectively, as follows:

$$Q = \overbrace{\begin{bmatrix} q_D & 0 & \cdots & 0 \\ 0 & q_D & 0 & 0 \\ \vdots & 0 & \ddots & \vdots \\ 0 & 0 & \cdots & q_D \end{bmatrix}}^{P\ times} \tag{13}$$

$$R = \overbrace{\begin{bmatrix} r_D & 0 & \cdots & 0 \\ 0 & r_D & 0 & 0 \\ \vdots & 0 & \ddots & \vdots \\ 0 & 0 & \cdots & r_D \end{bmatrix}}^{H\ times} \tag{14}$$

where $q_D \in \mathbb{R}^{m \times m}$ and $r_D \in \mathbb{R}^{w \times w}$ are diagonal matrices whose elements are derived from the output and input weight vectors, respectively. That is:

$$q_D = \begin{bmatrix} w_1 & 0 & \cdots & 0 \\ 0 & w_2 & 0 & 0 \\ \vdots & 0 & \ddots & \vdots \\ 0 & 0 & \cdots & w_m \end{bmatrix} \tag{15}$$

$$r_D = \begin{bmatrix} r_1 & 0 & \cdots & 0 \\ 0 & r_2 & 0 & 0 \\ \vdots & 0 & \ddots & \vdots \\ 0 & 0 & \cdots & r_w \end{bmatrix} \tag{16}$$

2.2.2 Reference trajectory vector

The reference trajectory can be specified in several different ways [2, 10]. An approach that specifies the reference trajectory as an exponential approximation from the current measurement $y(k)$ to the set point $s(k)$ is given as [17]:

$$r(k+j) = \alpha^j y(k) + \left(1 - \alpha_1^j\right) s_k \tag{17}$$

where α $(0 < \alpha \leq 1)$ is a filter on the output. Large value of α gives a slower, but smoother response.

Given that the current measurements of a plant with m output channels are represented as a vector $y = [y_1, y_2, \dots, y_{\mathrm{m}}]$, that the specified set points on the m output channels are represented as a vector $s = [s_1, s_2, \dots, s_{\mathrm{m}}]$, and that the filters on the output channels are specified as a vector $\alpha = [\alpha_1, \alpha_2, \dots, \alpha_{\mathrm{m}}]$, then the reference trajectory vector T $(T \in \mathbb{R}^{mP \times 1})$ over a prediction horizon P is defined as [17]:

$$T = [T_1, T_2, T_3, \dots, T_P]^T \tag{18}$$

where each T_i in T is defined as:

$$T_i = \left[\alpha_1^i y_1 + (1 - \alpha_1^i) s_i, \alpha_2^i y_2 + (1 - \alpha_2^i) s_2, \cdots, \alpha_m^i y_m + (1 - \alpha_m^i) s_m\right] \tag{19}$$

and α_j^i indicates output filter α for output channel j raised to power i.

If the vector of actual plant outputs over the prediction horizon, $Y \in \mathbb{R}^{mP \times 1}$ and set point vector over the same prediction horizon $S \in \mathbb{R}^{mP \times 1}$ are defined as:

$$Y = [\underbrace{y, y, \dots, y}_{P\ times}]^T \tag{20}$$

$$S = [\underbrace{s, s, \dots, s}_{P\ times}]^T \tag{21}$$

Then, the reference trajectory vector T for the entire prediction horizon P can be written compactly as:

$$T = \beta Y + \eta S \tag{22}$$

where both $\beta \in \mathbb{R}^{mP \times mP}$ and $\eta \in \mathbb{R}^{mP \times mP}$ are matrices derived from the output channels filters.

A diagonal matrix b is defined using the elements of the output filters vector α so that:

$$b = \begin{bmatrix} \alpha_1 & 0 & \cdots & 0 \\ 0 & \alpha_2 & \vdots & 0 \\ \vdots & \vdots & \ddots & \vdots \\ 0 & \cdots & 0 & \alpha_m \end{bmatrix} \tag{23}$$

Then, the matrix β is defined as a matrix of powers of b, that is,

$$\beta = \begin{bmatrix} b^1 & 0 & \cdots & 0 \\ 0 & b^2 & \vdots & 0 \\ \vdots & \vdots & \ddots & \vdots \\ 0 & \cdots & 0 & b^P \end{bmatrix} \tag{24}$$

The matrix η is then defined as:

$$\eta = I_{mP} - \beta \tag{25}$$

where I_{mP} is an identity matrix of the size mP.

2.3 The optimal control law

The objective function of Eq. (12) is stated as finding the smallest J that minimizes the squares of the errors between the reference trajectory vector T and the predicted output vector $\hat{Y}$ with consideration to the size of the input change vector ΔU. For linear unconstrained MPC, the optimal control law is obtained by finding the least squares solution of the quadratic objective function of Eq. (12). By substituting the prediction outputs Eqs. (7) and (8) into Eq. (12), we have:

$$\min_{\Delta U} J = \left(T - \Phi\Delta U - \Gamma\hat{X}\right)^T \mathrm{Q}\left(T - \Phi\Delta U - \Gamma\hat{X}\right) + (\Delta U)^T \mathrm{R}(\Delta U) \tag{26}$$

If a variable E is defined as:

$$E = T - \Gamma\hat{X} = \beta Y + \eta S - \Gamma\hat{X} \tag{27}$$

Substituting Eq. (27) into Eq. (26) gives:

$$\min_{\Delta U} J = (E - \Phi\Delta U)^T \mathrm{Q}(E - \Phi\Delta U) + (\Delta U)^T \mathrm{R}(\Delta U) \tag{28}$$

Expansion of Eq. (28) gives:

$$\min_{\Delta U} J = -2\Delta U^T \Phi^T \mathrm{Q}E + \Delta U^T \left(\Phi^T \mathrm{Q}\Phi + \mathrm{R}\right)\Delta U + E^T QE \tag{29}$$

The minimum of Eq. (29) occurs when its first derivative is zero. The first derivative of this equation with respect to ΔU equated to zero is:

$$\frac{\partial J}{\partial \Delta U} = 0 = -2\Phi^T QE + 2\left(\Phi^T \mathrm{Q}\Phi + \mathrm{R}\right)\Delta U \tag{30}$$

Then, the optimum control law for unconstrained MPC is found to be:

$$\Delta \breve{U} = \left(\Phi^T \mathrm{Q}\Phi + \mathrm{R}\right)^{-1}\Phi^T \mathrm{Q}E \tag{31}$$

or compactly as:

$$\Delta \breve{U} = \Psi E = \Psi\left(\beta Y + \eta S - \Gamma\hat{X}\right) \tag{32}$$

where

$$\Psi = \left(\Phi^T \mathrm{Q}\Phi + \mathrm{R}\right)^{-1}\Phi^T \mathrm{Q} = \Omega^{-1}\Phi^T \mathrm{Q} \tag{33}$$

and

$$\Omega = \Phi^T \mathrm{Q}\Phi + \mathrm{R} \tag{34}$$

For MPC as receding horizon control where only the first set of w inputs $\Delta \breve{u}(k)$ obtained in the control horizon is sent to the plant as $\Delta u(k)$, we can express $\Delta u(k)$ as:

$$\Delta u(k) = \Delta \breve{u}(k) = K_u \Delta \breve{U} = K_u \Psi\left(\beta Y + \eta S - \Gamma\hat{X}\right) \tag{35}$$

where $K_u \in \mathbb{R}^{w\times(\mathrm{H}-1)\mathrm{w}}$ is defined as:

$$K_u = \begin{bmatrix} I_w & 0_{w,(H-1)w} \end{bmatrix} \tag{36}$$

The matrix I_w is an identity matrix of size w, while $0_{w,(H-1)w}$ is an all-zeros matrix of w rows and $(H-1)w$ columns.

From Eq. (35), the optimum control law for the calculation of the first set of control input change $\Delta u(k)$ that is sent to the plant can be written in terms of constant matrices $K_y \in \mathbb{R}^{w\times \mathrm{mP}}$, $K_s \in \mathbb{R}^{w\times \mathrm{mP}}$, and $K_x \in \mathbb{R}^{w\times(\mathrm{n+m})}$ as:

$$\Delta u(k) = K_y Y + K_s S - K_x X \tag{37}$$

where

$$K_y = K_u \Psi \beta \tag{38}$$

$$K_s = K_u \ \Psi \eta \tag{39}$$

$$K_x = K_u \ \Psi \Gamma \tag{40}$$

It can be seen from the above definitions that the problem of MPC implementation can be divided into two parts. The first part involves computing the constant matrices K_y, K_s, and K_x. These matrices can be computed once the prediction horizon P, the control horizon H, the output weights vector q, the input weights vector r, and the output filters vector α are specified for a system of m outputs and w inputs. The second part involves the estimation of the vectors Y, S, and X at each sampling interval. The elements of Y are formed from immediate past plant measurements, while those of S are formed from specified set point values. The elements of X are obtained from measured or predicted states of the plant.

2.4 State-space MPC observer design

For state-space MPC, good quality control action requires good, distinct information about each of the states of the process to be controlled. However, not all the states of a process are always directly measurable. Some of the states are either too difficult or impractical to measure, or they occur as linear combinations of other states. In addition, there is always model-plant mismatch in reality.

To obtain good and real-time estimates of the augmented state vector X, state-space MPC usually incorporates a state estimator, a soft instrument generally referred to as an observer. The usual approach is to obtain estimates of the unknown states using information from the measured process outputs, inputs, and the dynamic models of the system. The common approach is the use of Kalman Filter.

2.4.1 Kalman filter and model predictive control

Kalman Filter is a very powerful estimator technique that has found widespread application in many fields. In the field of control, it is used popularly as a state estimator. The Kalman Filter technique was developed in the 1960s [33]. It uses measurements linearly related to the state but corrupted with white noise. It assumes that the discrete-time dynamic model of a system is stochastic, excited by random ("white") disturbances (or process noise) and random ("white") measurement noise [30, 34].

The Filter is based on the general form of the discrete-time stochastic dynamic model given in Eqs. (1) and (2). The Kalman Filter works correctly only when the system for which the states are to be estimated is *observable*. A given linear dynamic system model with a given linear input/output model is observable if and only if its state is uniquely determinable from the model definition, its inputs, and outputs [33].

Given, say, a linear dynamic system represented by the dynamic model of Eqs. (1) and (2) with n states, the system is observable only if:

$$rank\left(\begin{bmatrix} C_d & C_d A_d & C_d A_d^2 & \cdots & C_d A_d^{n-1} \end{bmatrix}^T \right) = n \tag{41}$$

The equivalent discrete-time stochastic dynamic system using the augmented state-space model that is used in the design of model predictive control is [11, 31]:

$$x(k+1) = Ax(k) + B_d\Delta u(k) + GZ(k) \tag{42}$$

$$y(k) = Cx(k) + V(k) \tag{43}$$

where $Z = \in \mathbb{R}^{(n+m)\times 1}$ is the new process noise vector, $V = \in \mathbb{R}^{m\times 1}$ is the new measurement noise vector, and $G = \in \mathbb{R}^{(n+m)\times(n+m)}$ is the new matrix relating the process noise to the state variables.

Since the process noise and measurement noise are assumed to be white, their respective covariance matrices are usually treated as diagonal matrices. The process noise and measurement noise covariance matrices are now defined, respectively, as $\mathrm{M} = \in \mathbb{R}^{(n+m)\times(n+m)}$ and $\mathrm{N} = \in \mathbb{R}^{m\times m}$.

There are currently many versions of the Kalman Filter, in continuous and discrete forms. The two common types used in linear MPC are the discrete predictor and the discrete predictor-corrector types. Their derivations are presented in the following sub-sections.

2.4.2 Predictor type Kalman filter

In the predictor type Kalman Filter, the state variables are estimated using the observer of the form [30]:

$$\hat{X}_p(k+1) = A\hat{X}_p(k) + B\Delta u(k) + K_p\big(y(k) - C\hat{X}_p(k)\big) \tag{44}$$

where gain $K_p = \in \mathbb{R}^{(n+m)\times m}$ is the predictor observer gain.

If $e_p(k)$ is the state estimation error, the state estimation error equation for the predicted type observer is given as:

$$e_p(k+1) = \big(A - CK_p\big)e_p(k) \tag{45}$$

The error will converge to zero if the eigenvalue of $\big(A - CK_p\big)$ is within the unit circle [11]. The optimal gain K_p that guarantees that the eigenvalues are inside the unit circle for the observable system is defined as [11]:

$$K_p = A\mathrm{W}C^T\big(C\mathrm{W}C^T + \mathrm{N}\big)^{-1} \tag{46}$$

where $\mathrm{W} = \in \mathbb{R}^{(n+m)\times(n+m)}$ is the auto covariance matrix of the predicted state estimate that satisfies the discrete-time algebraic Riccati equation [11]:

$$W(i+1) = A\mathrm{W}(\mathrm{i})A^T - K_pCW(i)A^T + N \tag{47}$$

By specifying an initial $\mathrm{W}(0)$, Eqs. (46) and (47) can be solved recursively to obtain the optimal K_p (the value at which K_p becomes constant). That is, using the initial $W(0)$, obtain a value of K_p using Eq. (46). Then use the new K_p and the old W to obtain a new W using Eq. (47). Repeat until a constant value of K_p is reached. The initial $W(0)$ can be prescribed as a matrix of the right dimension with guessed values. Common practice is to set $W(0)$ as an identity matrix $I_{nm} = \in \mathbb{R}^{(n+m)\times(n+m)}$.

Notice that in calculating the state estimate using Eq. (44), the past measurements $y(k)$ are used. For this case, there is a time delay of one-time step between the measurement y and the calculated current state estimate $\hat{X}(k+1)$. This can be a source of inaccuracy for some systems.

2.4.3 Predictor-corrector type Kalman filter

In the predictor-corrector type Kalman Filter, the state variables are estimated using the following two equations [30]:

$$\hat{X}_p(k+1) = A\hat{X}_c(k) + B\Delta u(k) \tag{48}$$

$$\hat{X}_c(k) = \hat{X}_p(k) + K_{pc}\left[y(k) - C\hat{X}_p(k)\right] \tag{49}$$

Eq. (48) is for the prediction of the states (predicted state estimates), while Eq. (49) is used to correct the prediction (corrected state estimates).

The state estimation error equation for the predicted-corrector type observer is given as:

$$e_c(k+1) = (A - CK_{pc}A)e_c(k) \tag{50}$$

The optimal gain $K_{pc} = \in \mathbb{R}^{(n+m)\times m}$ that ensures that the eigenvalues of $(A - CK_{pc}A)$ are inside the unit circle is defined as [30]:

$$K_{pc} = \mathrm{L}C^T\left(C\mathrm{L}C^T + \mathrm{N}\right)^{-1} \tag{51}$$

where $\mathrm{L} = \in \mathbb{R}^{(n+m)\times(n+m)}$ is the auto covariance matrix of the predicted state estimates defined as:

$$\mathrm{L} = AFA^T + G\mathrm{M}G^T \tag{52}$$

The duo of K_{pc} and L must satisfy the equation:

$$\mathrm{F} - \left[I - K_{pc}C\right]\mathrm{L} = 0 \tag{53}$$

where $\mathrm{F} = \in \mathbb{R}^{(n+m)\times(n+m)}$ is the auto covariance matrix of the corrected state estimate.

Eqs. (51)–(53) are usually solved recursively by specifying an initial L(0) (or initial F(0)). That is, using the initial $L(0)$ for example, obtain a value of K_{pc} using Eq. (51). Then use the new K_{pc} and the current L to obtain a value of F using Eq. (53). Use the new F to obtain a new L using Eq. (52). Repeat the process until a constant value of K_{pc} is reached. As before, the initial L(0) (or F(0)) may be set to a matrix of the right dimension with guessed values.

Note that for either the predictor type or the predictor-corrector type, the optimal gain (K_p or K_{pc}) is calculated a priori for the purpose of MPC applications.

3. Implementing steps for the linear unconstrained state-space MPC

In the implementation of MPC, the major tuning parameters are P, the prediction horizon, and M, the control horizon. Others are the output weight vector w, the input weight vector r, and the output filter vector τ. When these parameters are specified, the MPC constant matrices K_y, K_s, and K_x of Eq. (37), and either K_p or K_{pc} can be evaluated ahead. Then the vectors Y, S, and $\hat{X}$ have to be evaluated at every sampling time during the control process.

The steps involved in the implementations are outlined below. The outline is supported by an example of the control of a benchmark Wood-Berry model of a distillation column described in Ref. [17].

3.1 The distillation column model

The Wood-Berry distillation column model is a multivariable process with input delay times. The model is in step response format and given as [17]:

$$\begin{bmatrix} y_1 \\ y_2 \end{bmatrix} = \begin{bmatrix} \frac{12.8e^{-s}}{16.7s+1} & \frac{-18.9e^{-3s}}{21s+1} \\ \frac{6.6e^{-7s}}{10.9s+1} & \frac{-19.4e^{-3s}}{14.4s+1} \end{bmatrix} \begin{bmatrix} u_1 \\ u_2 \end{bmatrix} + \begin{bmatrix} \frac{3.8e^{-8.1s}}{14.9s+1} \\ \frac{4.9e^{-3.4s}}{13.2s+1} \end{bmatrix} d_1 \tag{54}$$

The variables to be controlled are the distillate composition y_1 and the bottom composition y_2. The two manipulated variables are the reflux flow rate u_1 and the steam flow rate u_2. One unmeasured disturbance d_1 is included.

Using the method contained in Ref. [23], a state-space equivalent of the model can be constructed. Application of the procedure shows that, with a sampling period of 5 minutes, a state-space model with seven states is required to have a perfect fit with the step response model. A state-space model with four states is not too far behind, as shown in **Figure 3**.

For the sake of compactness of presentation, the state-space model with four states is adopted for this example. The discrete-time four-state state-space model (with a sampling period of 5 minutes) is:

$$A_d = \begin{bmatrix} 0.8176 & -0.2432 & -0.01545 & 0.02131 \\ 0.2494 & 0.00755 & -0.01925 & -0.09369 \\ -0.01357 & -0.2665 & 0.5144 & 0.2699 \\ -0.01209 & -0.2709 & 0.07935 & 0.5491 \end{bmatrix}$$

$$B_d = \begin{bmatrix} -1.16 & 2.139 \\ 0.5479 & -1.464 \\ 1.258 & 0.4882 \\ -0.2957 & -0.1828 \end{bmatrix}$$

$$C_d = \begin{bmatrix} -1.712 & -0.9724 & 0.952 & -0.3119 \\ -1.725 & -1.169 & -1.079 & 0.07682 \end{bmatrix}$$

The MPC parameters adopted for the simulation are given in **Table 1**:

3.2 Calculation of the constant matrices K_y, K_s, K_x, and K_p or K_{pc}

The following steps may be taken to calculate the above constant matrices. The MATLAB codes for implementing the steps are contained in the attached Appendix A.

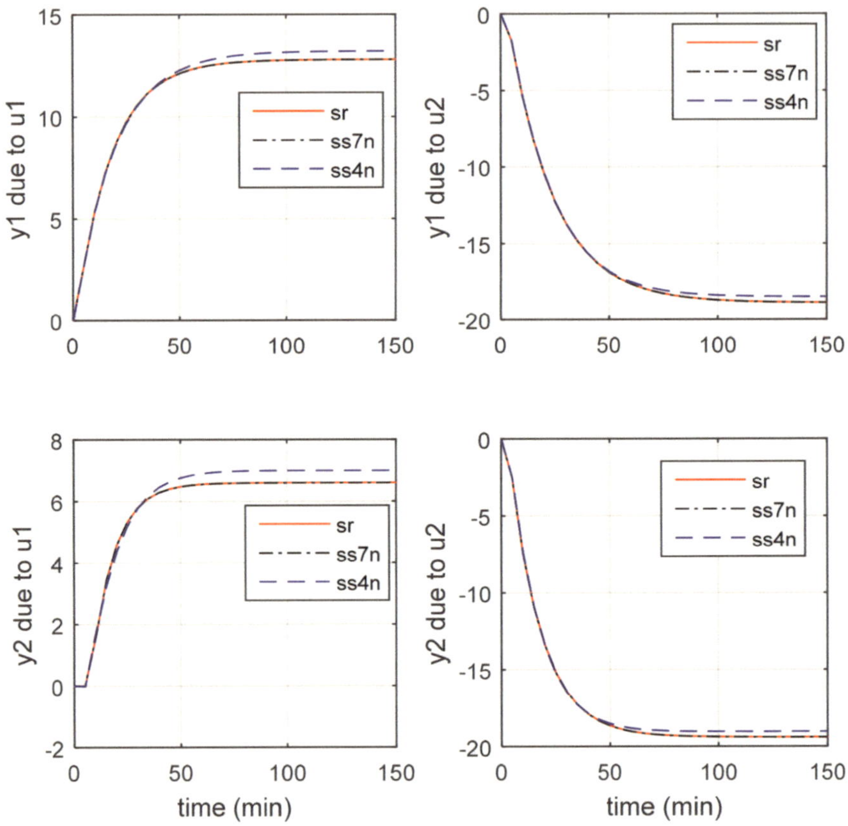

Figure 3.
Plot of the step response of the Wood-Berry model and Equivalent state-space models (sr = actual model; ss7n = seven states state-space equivalent model; ss4n = four states state-space equivalent model).

Prediction Horizon	***P***	**5**
Control Horizon	*H*	2
Output weights	*q*	[1, 1];
Input weight	*r*	[10, 10]
Output Filters	*α*	[0.9, 0.9]
Note: *P can be made much greater than five for better control. This value is chosen for the reduced computational load.*		

Table 1.
MPC parameters for the simulation.

Step 1:

Generate the augmented state matrices as given in Eqs. (3)–(6). These would be done using the discrete state-space matrices defined in Eqs. (1) and (2).

Step 2.

Generate the prediction matrices Φ and Γ as given in Eqs. (7) and (8). These would be done using the augmented state-space matrices generated in step 1.

Step 3.

Generate the output and input weight matrices Q and R of Eqs. (13) and (14). These would be done using the output weight vector q and input weight vector r.

Step 4.

Generate the reference trajectory matrix β defined in Eq. (24), and the matrix η defined in Eq. (23). These would be done using the output filter vector α.

Step 5.

Generate the constant matrix Ω as given in Eq. (34). Make use of the matrices from steps 2 and 3.

Step 6.

Evaluate the constant matrix Ψ as shown in Eq. (33). Make use of the matrices from steps 3 and 5.

Note that the matrix Ω is commonly non-square and rank deficient. Singular Value Decomposition (SVD) can be used for fabricating its pseudo-inverse in a way that numerical issues. Examples of the use of SVD for the solution of a set of linear equations, minimum-norm or least squares, for all the possible rank deficiencies of the coefficient matrix can be found in Ref. [35].

Step 7.

Evaluate constant matrices K_y, K_s, and K_x as shown in Eqs. (38)–(40). Make use of the value of Ψ evaluated in step 6.

Step 8.

Test for observability of the system (Eq. (41)).

Step 9.

Evaluate the observer gain. For Predictor type Kalman Filter, evaluate K_p algebraically using Eqs. (46) and (47). For predictor-corrector type Kalman Filter, evaluate K_{pc} algebraically using Eqs. (51)–(53).

Step 10.

Predefine constant vector V_P as a column vector of P ones, i.e., $V_P \in \mathbb{R}^{P\times 1}$

Predefining this vector will be useful during the periodic definition of vector Y (the measured vector for the prediction horizon) and vector S (the set point vector for the prediction horizon).

Step 11.

Calculate the eigenvalues of $\left(A - CK_p\right)$ or $\left(A - CK_{pc}A\right)$ to confirm that they are all inside the unit circle.

3.3 Periodic evaluation of the optimal control moves and MPC implementation

Step 1:

Run the MATLAB code given in Appendix A.

Step2:

Construct a Simulink file similar to that shown in **Figure 4** (for predictor only observer), or in **Figure 5** (for predictor-corrector type observer). In the figures, the blocks labeled A, B, C, K_s, K_y, K_x, and K_p/K_{pc} are Simulink gain blocks with their values set as the letters A, B, C, Ks, Ky, Kx, and Kp/Kpc used in the MATLAB code of Appendix A. The dotted blocks are MATLAB function blocks that implement the mathematical Kronecker tensor product $V_P \otimes y$ to give Y (defined in Eq. (20)) and of $V_P \otimes s$ to give S (defined in Eq. (21)). The block labeled *Plant* is the Simulink file for the Wood-Berry model given in Eq. (54).

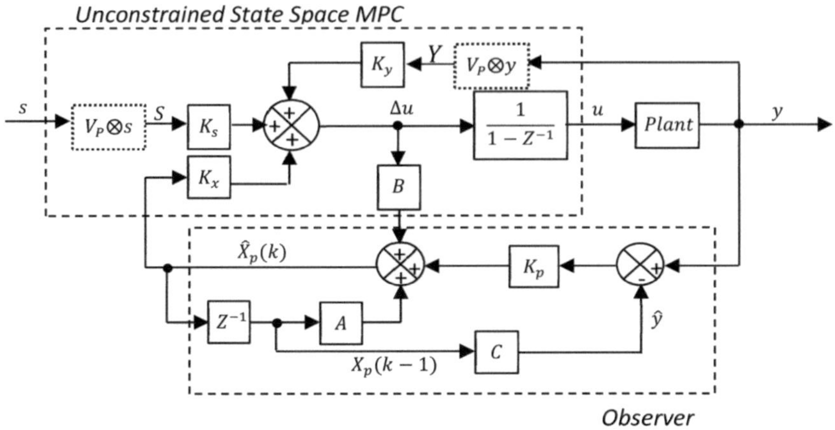

Figure 4.
Predictor type observer.

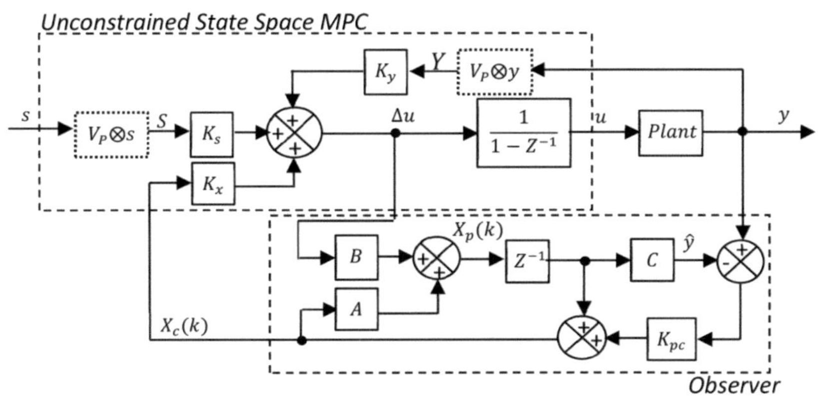

Figure 5.
Predictor–corrector type observer.

3.4 MPC implementation

When the MATLAB code of Appendix A is run, the constants K_s, K_y, K_x, and K_p/K_{pc} are found to be:

$$K_s = \begin{bmatrix} 0.0225, -0.0138, 0.0156, -0.0091, 0.0064, -0.0020, -0.0024, 0.0019, -0.0103, 0.0033 \\ 0.0069, -0.0116, -0.0045, -0.0212, -0.0043, -0.0112, -0.0004, 0.0017, 0.0054, 0.0144 \end{bmatrix}$$

$$K_y = \begin{bmatrix} 0.2023, -0.1239, 0.0666, -0.0389, 0.0173, -0.0052, -0.0046, 0.0037, -0.0149, 0.0048 \\ 0.0624, -0.1041, -0.0191, -0.0904, -0.0115, -0.0301, -0.0007, 0.0033, 0.0077, 0.0208 \end{bmatrix}$$

$$K_x = \begin{bmatrix} -0.1885 & -0.0581 & 0.2637 & 0.0942 & 0.2984 & -0.1791 \\ 0.4473 & -0.1941 & 0.1528 & 0.0413 & 0.0420 & -0.2284 \end{bmatrix}$$

$$K_p = \begin{bmatrix} -0.1217 & -0.1299 \\ -0.0540 & -0.0305 \\ 0.1276 & -0.1116 \\ 0.0901 & -0.0610 \\ 1.1821 & 0.2375 \\ 0.2150 & 1.2202 \end{bmatrix} K_{pc} = \begin{bmatrix} -0.1632 & -0.1744 \\ -0.0678 & -0.0321 \\ 0.1629 & -0.1734 \\ 0.1035 & -0.1057 \\ 0.8279 & 0.0726 \\ 0.0726 & 0.8448 \end{bmatrix}$$

The set point for y_1 is 0., and for y_2 is 1. A disturbance input d_1 with a step change from 0 to 0.3 is applied at t = 80 mins. The plots of the outputs and inputs are shown in **Figures 6** and 7, respectively.

In **Figure 6**, set point tracking without disturbance input for both controlled variables was achieved in about 50 minutes, which is approximately equal to the settling time of the plant (see **Figure 3**). After a disturbance input at 80 minutes, set point

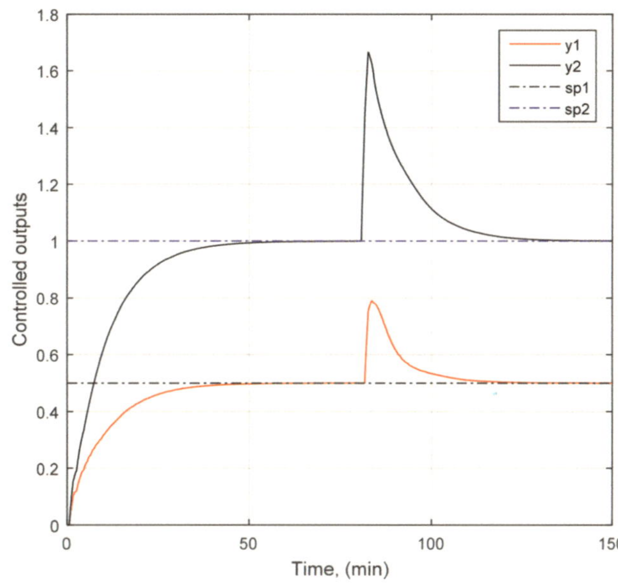

Figure 6.
Controlled output plots (sp1 = set point for y1; sp2 = set point for y2).

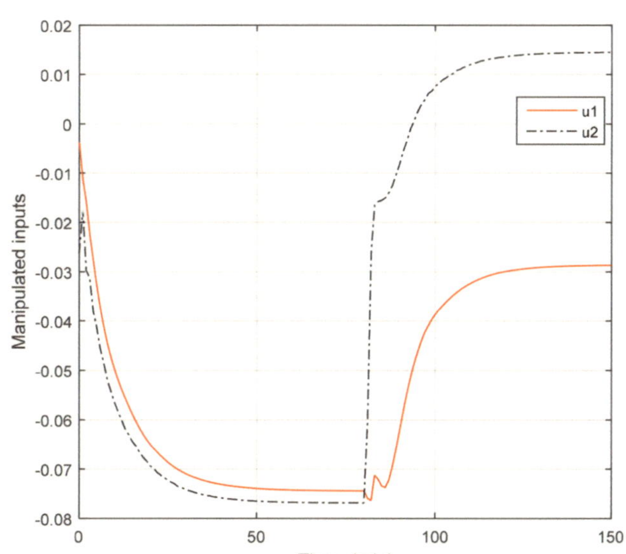

Figure 7.
Manipulated input plots.

tracking was achieved in less than 50 minutes later. These show that the MPC control's set point tracking and disturbance rejection are good.

4. Conclusions

In this chapter, a brief introduction of MPC generally, and linear, unconstrained state-space MPC is given. The linear quadratic objective function governing the formulation of MPC is presented and the optimal solution is discussed. The formulation of the prediction equation for state-space MPC is presented, and detailed consideration is given to the definition of the terms of the optimal control law. The design of two types of observers for the MPC includes the predictor only type and the predictor-corrector type. The steps for implementing the state-space MPC, together with its implementation on a benchmark plant model, are presented.

Conflict of interest

The author declares no conflict of interest.

Appendix A

```
clear
Ts = 5; %sampling interval

Ad =[0.8176    -0.2432     -0.01545    -0.02131
      0.2494    0.00755    -0.01925    -0.09369
      0.01357  -0.2665      0.5144      0.2699
     -0.01209  -0.2709      0.07935     0.5491];

Bd = [-1.16       2.139
       0.5479    -1.464
       1.258      0.4882
      -0.2957    -0.1828];

Cd = [-1.712    -0.9724    0.952    -0.3119
      -1.725    -1.169    -1.079     0.07682];

Dd = zeros(size(Cd,1),size(Bd,2));

% MPC tuning parameters
P = 5;% prediction horizon
H = 2; % control horizon
q = [1,1];%outputs weights
r = [1,1]; %input weights
sALPHA = [0.9,0.9]; % filters on outputs

%derivided simulation parameters
n = size(Ad,1);% no of states
m = size(Cd,1);% no of outputs
w = size(Bd,2);% no of inputs
ng = n+m; % no of augmented states
```

```
%%
%>>>Part 1

%step 1: generate MPC augmented state matrices
Onm = zeros(n,m);
Omn = Onm';
Im = eye(m);
A = [Ad,Onm;Cd*Ad,Im]; %augmented matrix A
B = [Bd;Cd*Bd];% augmented matrix B, Eq 3
C = [Omn,Im];% augmented matrix C, Eq 4
D = zeros(size(C,1),size(B,2));
%%
%step 2: generate prediction matrices for the prediction
equation
% Uppercase gama (bGAMA) using Eq 5
bGAMA = zeros(m*P,ng);
for i = 1:P
    bGAMA((i-1)*m+1:i*m,:) = C*(A^i);
end
% Uppercase phi (bPHI) using Eq 5
PHI1 = zeros(m*P,w);
for i = 1:P
   PHI1((i-1)*m+1:i*m,:) = C*(A^(i-1))*B;
end
bPHI = zeros(m*P,w*H);
for j = 1:H
    bPHI((j-1)*m+1:end,(j-1)*w+1:j*w) = PHI1(1:end-(j-1)*m,:);
%Eq 5
end
%%
%step 3: generate output and input weight matrices
qD = diag(q);%diagonal matrix of the output weights
rD = diag(r);% diagonal matrix of the input weights
Q = kron(eye(P),qD);%output penalty matrix, Eq 10
R = kron(eye(H),rD);%input penalty matrix, Eq 11
%%
%step 4: generate reference trajectory matrices
%lower case beta (sBETA), Eq 19
b = diag(sALPHA);
sBETA = zeros(m*P,m*P);
for i = 1:P
    sBETA((i-1)*m+1:i*m,(i-1)*m+1:i*m) = b^i;
end
%lower case eta (sETA)
sETA = eye(m*P) - sBETA;% Eq 20
%%
%step 5: generate uppercase omega (bOMEGA)
bOMEGA = (bPHI'*Q*bPHI + R);%Eq 28
%%
%step 6: generate uppercase psi (bPSI)
bPSI = pinv(bOMEGA)*(bPHI'*Q);%Eq 27
%%
%step 7: Evaluate the control law constant matrices
Ku = [eye(w),zeros(w,(H-1)*w)];%Eq 28
Ky = Ku*bPSI*sBETA; %Eq 30
Ks = Ku*bPSI*sETA; %Eq 31
```

```
Kx = Ku*bPSI*bGAMA;% Eq 32
%%
%step 8: Test for observability, Eq 35
if rank(obsv(A,C)) == ng
    disp('system is observable')

%Step 9: Calcuklate the observer gains
    M = 0.1*eye(ng);%process noise covariance matrix
    N = 0.1*eye(m);%measurement noise covariance matrix
    G = eye(ng);%matrix relating the process noise to the state
variables
    mnit = 1000;% maximum number of iterations

%A) Predictor type observer
% solving Eqs (40)-(41) recursively
    W0 = eye(ng);%initial autocovariance matrix of predictor
state estimate
    W = W0;
    Kpo = (A*W*C')/(C*W*C' + N);     %Eq 40
    for k = 1:mnit
        W = A*W*A' + M - (A*W*C')*Kpo';    %Eq 41
        Kp = (A*W*C')/(C*W*C' + N);%Eq 40
        nm = norm(Kpo-Kp);
        if nm <= 1e-10;
            break
        end
        Kpo = Kp;
    end

%B) Predictor -corrector Type
    L0 = 0.01*eye(ng);%initial autocovariance matrix of
predictor/corrector
                   %state estimate
    In = eye(ng);
    L = L0;
    Kpco = (L*C')/(C*L*C' + N); %Eq 45
    for k = 1:mnit
        F = (In - Kpco*C)*L; %Eq 47
        L = A*F*A' + G*M*G'; %Eq 46
        Kpc = (L*C')/(C*L*C' + N); %Eq 45
        nm = norm(Kpco-Kpc);
        if nm <= 1e-10;
            break
        end
        Kpco = Kpc;
    end
else
   error('system no observable')
end
%%
%step 10: Predefine constant matrix for periodic calculations
 VP = ones(P,1);

%step 11: check that eigenvalues of (A-Kp*C) and (A-Kpc*C*A) are
all
%inside the unit circle
[~,el1] = eig(A-Kp*C);
ell1 = diag(el1);%eigenvalues of (A-Kp*C) as column vector
[~,el2]=eig(A-Kpc*C*A);
ell2 = diag(el2); %eigenvalues of (A-Kpc*C*A) as column vector
```

Author details

Mohammed Tajudeen Jimoh
Department of Mechanical Engineering, Bayero University, Kano, Nigeria

*Address all correspondence to: mtjimoh.mec@buk.edu.ng

References

[1] Qin SJ, Badgwell TA. A survey of industrial model predictive control technology. Control Engineering Practice. 2003;**11**:733-764

[2] Camacho EF, Bordons C. Model Predictive Control. Verlag, London: Springer; 2007

[3] Charos GN, Arkun Y, Taylor R. Model predictive control of an industrial lime kiln. Tappi Journal. 1991;**74**:203-211

[4] Richalet J. Industrial applications of model based predictive control. Automatica. 1993;**29**(5):1251-1274

[5] Zanovello R, Budman H. Model predictive control with soft constraints with application to lime kiln control. Computers & Chemical Engineering. 1999;**23**:791-806

[6] K. Alexis, N. G and T. A., Model predictive quadrotor control: attitude, altitude and position experimental studies, IET Control Theory and Applications, 6. 1-16. 2014

[7] Utku E, Prach A, Koçer BB, Raković VS, Kayacan E, Açıkmeşe B. Model predictive control in aerospace systems: Current state and opportunities. Journal of Guidance, Control and Dynamic. 2017;**40**(7):1541-1566

[8] Gold T, Volz A, Graichen K. Model predictive interraction control for industrial robot. IFAC-Papers Online. 2020;**53**(2):9891-9898

[9] Hugo A. Limitations of model predictive controllers. Hydrocarbon Processing. 2000;**79**(1):83-88

[10] Maciejowski JM. Predictive Control: With Constraints. Essex, England: Pearson Education; 2002

[11] Wang L. Model Predictive Control System Design and Implementation Using MATLAB®. Verlag, London: Springer-Verlag; 2009

[12] Bemporad A, Ricker NL, Morari M. Model Predictive Control Toolbox User's Guide. Natick, Massachusetts: The Mathworks, Inc.; 2020

[13] Kwon W, Han S. Receding Horizon Control Model Predictive Control for State Models. Berlin: Springer; 2005

[14] Curtler CR, Ramaker BL. Dynamic matrix control – a computer control algorithm. In: Joint Automatic Control Conference. Vol. 1, No. 17. San Francisco, California; 13-15 Aug 1980. p. 72

[15] Hokanson DA, Gerstle JG. Dynamic matrix control multivariable controllers. In: Practical Distillation Control. New York: Springer; 1993

[16] Huang B, Kadali R. Model predictive control: Conventional approach. In: Dynamic Modeling, Predictive Control and Performance Monitoring. London: Springer; 2008

[17] Seborg DE, Edgar TF, Mellichamp DA, Doyle FJ. Process Dynamics and Control. 4th ed. New Jersey, USA: John Wiley & Sons; 2016

[18] Clarke DW, Mohtadi C, Tuffs P. Generalized predictive control—Part I. The basic algorithm. Automatica. 1987; **23**:137-148

[19] Camacho E. Constrained generalized predictive control. Automatic Control, IEEE Transactions. 1993;**38**:327-332

[20] Marquis P, Broustail JP. SMOC, a bridge between state space and model

predictive controllers: Application to the automation of a hydrotreating unit. IFAC Proceedings. 1988;**21**(4):37-45

[21] Yousfi C, Tournier R. Steady state optimization inside model predictive control. In: American Control Conference. Boston, MA, USA: IEEE; 1991

[22] Rawlings JB. Tutorial: Model predictive control technology. In: Proceedings of America Control Conference, San Diego, California. June 1999. pp. 662-676

[23] Jimoh MT, Dan'Isa A. State space model realization using step response data of MIMO system with input delays for model predictive control. ActPeriodica Technologica, APTEFF. 2019;**50**:93-104

[24] Garcia CE, Prett DM, Morari M. Model predictive control: Theory and practice—A survey. Automatica. 1989; **25**:335-348

[25] Muske KR, Rawlings JB. Model predictive control with linear models. AICHE Journal. 1993;**39**:262-287

[26] Kothare MV, Balakrishnan V, Morari M. Robust constrained model predictive control using linear matrix inequalities. Automatica. 1996;**32**:1361-1379

[27] Borrelli F, Baotić M, Bemporad A, Morari M. Dynamic programming for constrained optimal control of discrete-time linear hybrid systems. Automatica. 2005;**41**:1709-1721

[28] Ricker NL. Model predictive control: State of the art. Proceedings of the Chemical Process Control. 1991;**4**: 271-296

[29] Rawlings JB. Tutorial overview of model predictive control. Control Systems, IEEE. 2000;**20**:38-52

[30] Haugen F. State estimation with Kalman Filter [internet]. In: Kompendium for Kyb. 2. Norway: Høgskolen i Oslo; 2016. pp. 101-127. Available from: http://techteach.no/fag/seky3322/0708/kalmanflter/kalmanfilter.pdf

[31] Roset B, Nijmeijer H. Observer based model predictive control. International Journal of Control. 2004;**77**(17): 1452-1462

[32] Li S, Lim KY, Fisher DG. A state space formulation for model predictive control. AICHE Journal. 1989;**35**(2): 241-249

[33] Grewal MS, Andrews AP. Kalman Filtering: Theory and Practice Using MATLAB. 2nd ed. New York: John Wiley and Sons Inc.; 2001

[34] Kim Y, Bang H. Introduction to Kalman filter and its applications [internet]. In: Introduction and Implementation of the Kalman Filter. London, UK: IntechOpen; 2019. DOI: 10.5772/intechopen.80600

[35] Yang WY, Cao W, Chung T-S, Morris J. Applied Numerical Methods Using Matlab. New Jersey: John Wiley & Sons, Inc., Publication; 2005

Chapter 3

Robust Control of Autonomous Robotic Vehicles Using Nonlinear Model Predictive Controller

Aliasghar Arab

Abstract

The proposed chapter will explore the application of robust control methods for electric connected autonomous robotic vehicles (EACVs) with a focus on robust control techniques. As the demand for autonomous mobility increases, ensuring their safety, reliability, and efficiency becomes paramount. This chapter will delve into the challenges and solutions associated with the robust control of EACVs, particularly under varying operational conditions that cause uncertainties. By leveraging robust and learning control strategies, the chapter aims to demonstrate how these methods can enhance the performance and resilience of EACVs. Key topics will include the integration of nonlinear control for trajectory following, handling disturbances and parameter variations, and ensuring system robustness. The chapter will also present case studies and simulations to illustrate the effectiveness of these control strategies in real-world scenarios. This comprehensive overview will provide valuable insights for researchers, engineers, and practitioners involved in the development and deployment of autonomous robotic vehicle technologies.

Keywords: robust control, autonomous vehicles, mobile robots, nonholonomic systems, nonlinear control

1. Introduction

There is increasing focus on employing autonomous robotic vehicles across various sectors, including robot taxis for transportation, delivery trucks and robots for logistics, and autonomous mobile platforms for industrial automation. These intelligent robotic systems are particularly valuable in hazardous environments where human operation poses safety risks. Notably, wheeled nonholonomic robotic vehicles are highly regarded for such applications. As future mobility trends shift toward connected and electric-powered systems, control strategies that directly manage voltage will become increasingly crucial for these advanced autonomous vehicles.

Globally, approximately 94% of the 1.25 million annual automotive fatalities—including 40,000 deaths in the United States—are attributed to driver behavior. The World Health Organization reports that 49% of road traffic deaths involve pedestrians (22%), cyclists (4%), and motorcyclists (23%), as well as vehicle collisions with animals and accidents related to natural disasters [1]. The significant role of human drivers in these incidents is unlikely to diminish without technological and scientific advancements that make autonomous vehicles (AVs) more reliable than human drivers [2]. Much like Anti-Lock Braking Systems (ABS) and Electronic Stability Control (ESC), AVs are designed to reduce fatalities and serious injuries among vulnerable road users by enhancing vehicle safety during emergency situations [3]. Nevertheless, to meet the safety and reliability standards accepted by the industry, advanced control methods are essential to further improving road safety.

A key challenge in controlling connected, electric-powered, mechanically constrained systems—known as nonholonomic robotic vehicles—lies in addressing various model and environmental uncertainties. These uncertainties can arise from factors such as inaccuracies in kinematic and dynamic models, external disturbances from unknown environments, fluctuating reference trajectories, and errors in measurement noise and perception. To manage these challenges, several classical control techniques have been proposed, including $H\infty$ control [4], sliding mode control [5], adaptive control [6], adaptive sliding mode control [7], observer-based feedback linearization control [8], nonlinear model predictive control [9], and robust nonlinear control [10]. Simultaneously, intelligent control approaches have emerged to handle complex systems, leveraging fuzzy logic, neural networks, and other learning algorithms to approximate functions or derive rules based on expert knowledge. Notable examples include adaptive learning control [11], adaptive fuzzy control [12], fuzzy neural control method [13], and fuzzy controller enhanced using evolutionary optimizations [14, 15].

Ensuring effective motion and tracking control in electric-powered autonomous robotic vehicles relies on managing the voltage and current of electric motors to generate the necessary forces and torques [16]. Achieving precise control in these systems requires addressing practical limitations, such as motor voltage and current constraints imposed by battery-powered systems, especially during high-speed, high-precision maneuvers. Incorporating the dynamics of the motors into the control framework is crucial for enhancing accuracy and reliability. Trajectory tracking methods that explicitly account for motor dynamics within the control structure have demonstrated success in addressing uncertainties [17, 18]. This integration of actuator dynamics has emerged as a promising alternative to traditional multi-layer control approaches [19].

In contrast to torque control—which is nonlinear, coupled, and computationally demanding due to the complexity of robotic dynamics—voltage control simplifies the control process [20]. By using motor voltage as the direct control input, systems can achieve quicker, more robust performance with minimal tracking errors. This method not only improves control efficiency but also directly addresses the electrical characteristics of the motors, making it a key factor in controlling electric-powered autonomous robotic vehicles. Precise voltage control is essential for ensuring accurate motion and reliable trajectory tracking in these vehicles.

In this chapter, we model an electric robotic vehicle with electric motors in a state-space framework and design a robust voltage controller for trajectory tracking in autonomous robotic vehicles. This approach eliminates the need for a separate torque control loop, instead directly controlling motor voltages, simplifying the system's control architecture. The proposed control law proves highly effective in managing uncertainties, offering a fast, robust response with minimal tracking error. Through both simulations and experimental results, the voltage control approach is compared to traditional methods, such as adaptive feedback linearization and PID control, demonstrating its superior performance.

2. Modeling an electric robotic vehicles

A robotic vehicle powered by electricity can be modeled through kinematics, dynamics, and actuator modeling. Kinematics focuses on the geometric relationships of motion without accounting for forces. Dynamics examines how motion is influenced by forces, while actuator modeling connects the control input to the robot's torque input. Take, for example, a nonholonomic robotic vehicle with four wheels, each independently driven by electric motors. Each wheel is powered by an electric motor, as shown in **Figure 1**. The model is represented using $OXYZ$ as the global coordination reference and $O_rX_rY_rZ_r$ as a local coordination reference placed at the center of gravity on the robotic vehicle. The position $C_G \cong O_r$ is at the center of mass of the robotic vehicle. The length L is the distance from the axis crossing the center of the rear wheel to the center of the front wheel when the wheels are not steered. The length W is the distance between the right and left wheels, and d is the diameter of the wheel.

The position of the robotic vehicle in a vector format $q = [x \ y \ \theta]^T$, in which $[x \ \ y]$ is the position of O_r in the global reference frame, and θ is the heading direction taken counterclockwise from the axis of local reference frame X_r in respect to the axis of the global frame X. Any robotic vehicle can be represented by kinematic, dynamic, and actuator models. Depending on the application and the level of control precision required for the robotic vehicle's operation within its specific domain, these models can be refined to more detailed levels. Additionally, external factors such as sensors, surrounding objects like other vehicles, vulnerable road users, and environmental conditions like weather and lighting play a crucial role in the control of autonomous vehicles. This chapter focuses on designing a robust controller based on the vehicle's model, assuming that sensor data is reliable within a certain range.

2.1 Kinematics of a robotic vehicles

Kinematics of a robotic vehicle is very important for understanding the relationship between wheel rotational velocities $\acute{\varphi} = [\ \acute{\varphi}_{fr} \ \acute{\varphi}_{fl} \ \acute{\varphi}_{rr} \ \acute{\varphi}_{rl}\ ]$ and longitude velocities of

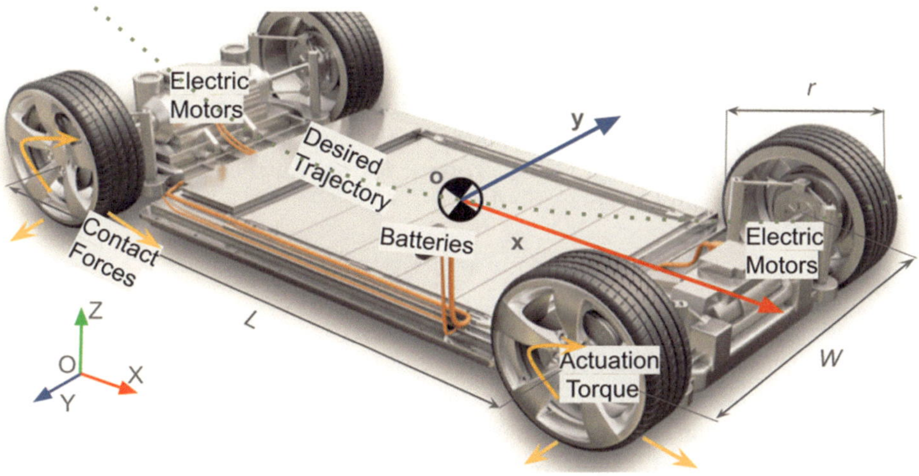

Figure 1.
Schematic view of an electric autonomous robotic vehicle.

the wheels $v_w^x = \left[v_{wfr}^x \;\; v_{wfl}^x \;\; v_{wrr}^x \;\; v_{wrl}^x \right]$ and steering angles $\delta = \left[\delta_{fr} \;\; \delta_{fl} \;\; \delta_{rr} \;\; \delta_{rl} \right]$. These vectors represent the states of the front-right, front-left, rear-right, and rear-left wheels, respectively. To extract the Kinematic equations, we need to understand the relationship between the wheel's longitude velocity v_w^x and wheel's rotational velocity $\acute{\varphi}$ as

$$v_w^x = \frac{r_w}{2}\acute{\varphi} + \sigma, \tag{1}$$

where σ is a vector to represent the slip ratio of the wheels. In an ideal situation, the slip of the wheels should be zero, hence the robot velocity vector $v = \left[v_x \;\; v_y \;\; \omega \right]^T$, where v_x, v_y and ω are the longitudinal, lateral, and angular velocity of a robotic vehicle, is governed by the drivetrain of kinematics.

$$\acute{\varphi} = T_w^\alpha(\delta)v + \hat{\sigma} \tag{2}$$

where

$$T_w^\alpha(\delta) = \tag{3}$$

The inverse-kinematic solution of the robotic vehicle can be obtained by setting the skidding of the robotic vehicle to be negligible. A small skidding value of a wheeled robotic vehicle means the lateral velocity of the wheels is equal to zero $v_w^y =$. Anti-lock braking system (ABS) and Electronic Stability Control (ESC) systems in the vehicles are designed to use local feedbacks to avoid potential slipping and skidding in the vehicles. Without slipping or skidding, the kinematic solution can be integrated with local feedback controllers to maintain the desired wheel angle and velocity. But in real world, slipping and skidding always exist and using kinematic models for precise control design is insufficient.

2.2 Dynamics of a robotic vehicles

However, slipping and skidding are inevitable and they will cause uncertainty and affect the controller's performance in an autonomous vehicle to follow a desired path. Also, the effects of vehicle's dynamic modeling are necessary for designing a robust and reliable control for an autonomous robotic vehicle. There are several methods for representing a vehicle's dynamics, including a simplified bike model based on the vehicle's symmetry [21], a four-wheel model with front steering [9], and a four-independent-wheel model [22]. Using a robotic format for the equations of motions for a robotic vehicle can be represented as

$$M(q)\acute{v} + C(q,v)v + \tau_d + A^T(q)\lambda = B(q)\tau \tag{4}$$

where $M(q) \in R^{3\times3}$, the inertia matrix, is symmetric and positive definitive, $C(q,v)v \in R^3$ is centrifugal and Coriolis forces and torques vector. Where the gravitational effects, frictions, and slip and skidding effects are considered as lumped uncertain disturbance vector $\tau_d \in R^3$. The non-square matrix $B(q) \in R^{3\times2}$ transforms the input torques from the joint space to the workspace. The vector $A(q) \in R^3$ is associated with constraints, and $\lambda \in R$ is the constraint force.

Assumption 1: The wheels do not slide under the nonholonomic constraint; hence, the lateral speed will stay zero.

$$v_w^y = 0 \tag{5}$$

Assumption 2: All kinematic equality constraints are time-independent so that

$$A(q)\acute{q} = 0 \tag{6}$$

Consider the velocity vector denoted as $v = \begin{bmatrix} v_x & v_y & \omega \end{bmatrix}^T$ where v represents the linear velocities in x and y directions and ω is angular velocity in the reference frame under the nonholonomic constraint, arising from pure rolling without slipping. Therefore,

$$v = \acute{T}_w^x(\delta)\acute{\varphi} \tag{7}$$

2.3 Modeling the actuator in a robotic vehicle

The electric motors provide the torques to the wheels *via* the following dynamics

$$J_m\acute{\varphi}_m + B_m\acute{\varphi}_m + r_g\tau_w = \tau_m \tag{8}$$

where $\tau_m = \begin{bmatrix} \tau_{mfr} & \tau_{mfl} & \tau_{mrr} & \tau_{mrl} \end{bmatrix}$ is the motor torque vector and $\tau_w = \begin{bmatrix} \tau_{wfr} & \tau_{wfl} & \tau_{wrr} & \tau_{wrl} \end{bmatrix}$ is the wheel torque vector, J_m, B_m, and r_g are the motor coefficients, namely the inertia, damping, and reduction gear, respectively. The motor velocities $\acute{\varphi}_m = \begin{bmatrix} \acute{\varphi}_{mfr} & \acute{\varphi}_{mfl} & \acute{\varphi}_{mrr} & \acute{\varphi}_{mrl} \end{bmatrix}$ and wheel velocities $\acute{\varphi}$ in Eq. (2) are related through the gears as

$$\acute{\varphi}_m = \frac{1}{r_g}\acute{\varphi} \tag{9}$$

Substituting (4) into (8), and using (9) yields to

$$\frac{J_m}{r_g}\frac{d\left(T_w^x(\delta)v + \hat{\sigma}\right)}{dt} + \frac{B_m}{r_g}\left(T_w^x(\delta)v + \hat{\sigma}\right) + r_g\tau_w = \tau_m \tag{10}$$

The differential equation of geared electric motors is

$$L_m\acute{I} + R_mI + \frac{K_b}{r_g}\acute{\varphi} = V \tag{11}$$

where $V = \begin{bmatrix} V_{fr} & V_{fl} & V_{rr} & V_{rl} \end{bmatrix}$ is the motor voltage vector and $I = \begin{bmatrix} I_{fr} & I_{fl} & I_{rr} & I_{rl} \end{bmatrix}$ is the vector of motor currents. R_m, L_m, K_b represent the coefficients of armature resistance, inductance, and back-emf constant, respectively. The motor torque vector τ_m is produced by the motor current vector

$$\tau_m = K_mI \tag{12}$$

$K_m \cong K_b$ is the torque constant based on the motor's specifications. Substituting Eqs. (12), (4), and (10) into (11) yields

$$L_m\acute{I} + R_m I + \frac{K_b}{r_g} T_w^{\alpha}(\delta)(v + \hat{\sigma}) = V \tag{13}$$

where $\hat{\sigma}$ captures the slip, sliding, and all modeling uncertainty effects, which should be addressed for a robust control design.

2.4 Electric robotic vehicle's state-space model

Using (3)–(13), the state-space model of the electric robotic vehicle system is then derived as

$$\acute{z} = f(z, u) + \hat{f}(z, u, \acute{\rho}) \tag{14}$$

where the state vector $z^T = \begin{bmatrix} q^T & v^T & I^T \end{bmatrix}$ and the input vector $u^T = \begin{bmatrix} \delta^T & V^T \end{bmatrix}$ and the nominal part of the nonlinear model of the system $f(z, u)$ is time-invariant and is extracted analytically using the dynamic properties of the system. An estimated model can be used for the uncertain part of the model $\hat{f}(z, V, \acute{\rho})$ as a lumped uncertainty approximation function. However, this function is uncertain, the range of uncertainties is limited, and we are assuming the error in estimating value of slip angle is $\|\rho - \acute{\rho}\|$. The nonlinear function, $\hat{f}$ is for compensating the dynamic parameter variations, external disturbances, and noises with upper and lower limit as $\left\|\hat{f}(z, u, \acute{\rho})\right\| < \gamma$. Accurate modeling is necessary to predict the state of the robotic vehicle for the receding horizon. However, precise model of the system f is not always available, and the estimation of $\hat{f}$ might be used for improving the robustness. Combining (4)–(14), the $f(z, u)$ term expands into

$$\acute{q} = \begin{bmatrix} cos\theta & -sin\theta & 0 \\ sin\theta & cos\theta & 0 \\ 0 & 0 & 1 \end{bmatrix} v,$$

$$\acute{v} = -M(q)^{-1}\left(C(v)v + \frac{K_b}{r_g} T_w^{\alpha}(\delta) I\right), \tag{15}$$

$$\acute{I} = \frac{V - RI - \frac{K_b}{r_g r_w} T_w^{\alpha}(\delta) v}{L_m}.$$

The state-space equations in (15) describe the model of a robotic vehicle as a coupled nonlinear multivariable system. The complexity of this model is a well-known challenge in the modeling and control of mobile robots and robotic vehicles.

3. Robust predictive control design

The fundamental idea is to design a robust predictive controller such that the robotic vehicle predicts the future states and incorporates the unknown part of the model into the prediction horizon to compensate for the uncertainties. Robustness constraints can then be incorporated as a cost function to constrained optimization of Nonlinear Model Predictive Controller (NMPC). The focus is on robustly tracking a desired trajectory with a wheeled robotic autonomous vehicle, so we design the controller in the context of the four-wheeled vehicle platform with four independent drive motors and no steering. Hence, the problem can be reduced to a model of nonholonomic differential wheel robotic vehicle platform. However, the challenge of not having a slip or sliding effect will add a burden to model-based controllers, such as NMPC.

Given that an upper bound of the system uncertainty is known, the controller employs iterative optimization to estimate the robotic vehicle's states over the prediction horizon, assuming zero steering wheel angles and specified drive wheel voltages. A modified version of nonlinear conjugate gradient (NCG) optimization algorithm is used [22]. The tracking error of a robotic vehicle is defined as $e = \|q_d - q\|$ for a given waypoint on the desired trajectory q_d. Initial guess for NCF optimization method is calculated using the output of the feedback linearization controller. Hence, NCG optimization in nonlinear predictive controller minimizes the cost function subjected to input and state constraints like robustness constraints as

$$min_{u(t)} \int_{\acute{t}=t}^{\acute{t}=t+t_p} \left(\alpha_1 e(\acute{t})^2 + \alpha_2 \acute{e}(\acute{t})^2 + J(\acute{\rho}, \acute{t})\right) d\acute{t}$$

subj. eq. (14) and

$$\begin{gathered} \left\|\hat{f}(z, u, \acute{\rho})\right\| < \gamma \\ u_{min} \le u(t) \le u_{max} \\ \delta(t) = 0 \end{gathered} \tag{16}$$

where $e(t+\rho)$ denotes the predicted error at ρ ahead on prediction horizon at time instant t and t_p denotes the prediction horizon. Matrices α_1 and α_2 are weighting parameters for error and the derivative of the error. The parameters $J(\acute{\rho}, \acute{t}) > 0$ is used for penalizing the robustness term if the system is violating any robustness factors. NMPC controller in Eq. (16) solves a constrained optimization problem to provide optimal control commands. In particular, the robustness requirements add constraints of the form $\left\|\hat{f}(z, u, \acute{\rho})\right\| < \gamma$. Lipschitz continuity of the cost functions is required for satisfying all the constraints in (16) at each episode of the predictive controls. The constrained optimization penalizes the cost function if the robotic vehicles tend to drift from the robust zone. Hence, the NCG optimizer in Eq. (16) will find the input vector for the robot, which will move the robotic vehicle into a safe zone, such that $\left\|\hat{f}(z, u, \rho)\right\| \to 0$. If the target is pushing the vehicle in unsafe state, the solver will stay at a local minimum, and the robot will send infeasible target messages to the planner to recalculate the desired trajectory.

4. Programming a nonlinear model predictive controller

Programming a Nonlinear Model Predictive Controller (NMPC) with code is essential for deepening the understanding of the mathematical, physical, and control system principles involved. By implementing NMPC algorithms in code, one can directly engage with the mathematical formulations, such as system dynamics and constraints, and see how these equations translate into real-world control actions. It allows for hands-on interaction with the physics of the system, such as handling nonlinearity, disturbances, and model predictions over time. Furthermore, coding the controller provides insight into how optimization techniques are applied to solve complex control problems in real time, helping bridge the gap between theoretical concepts and their practical applications in systems like autonomous vehicles or robotics.

There are numerous ways to simulate such a controller and validate the proposed controller in this chapter. Our suggestion is to try to formulate the problem using MATLAB m-files or Python to understand each step. Using out of shelf solvers or physics engines to simulate the vehicle model or implement NMPC is also an option but not recommended. An example of a U-turn trajectory following high-speed autonomous robotic race vehicles can be found here.

5. Validations through simulations

The proposed motion controller is implemented using numerical simulations in MATLAB, and a dynamic model in Carla is built, incorporating all essential physical properties based on a real robotic vehicle. The controller operates with nominal values for physical parameters, allowing a 20% tolerance. In the simulation, we account for a finite slip effect during various time periods, with no slip outside this window. The robust control parameters are tuned using heuristic optimizations. Validations through simulations are a critical step in the development and deployment of autonomous vehicles for several key reasons:

1. Safety: Autonomous vehicles must operate in a wide range of scenarios, including those that involve unexpected events or disturbances. Testing in real-world conditions without proper validation can pose significant risks. Simulations allow for the safe testing of the vehicle's control systems and algorithms without risking human life or damaging the vehicle.

2. Scenario diversity: Autonomous vehicles are expected to handle various complex maneuvers, such as lane changes, merging, and U-turns, under different environmental conditions. Simulations provide a controlled environment where these different scenarios can be replicated precisely and consistently. This allows developers to study how the vehicle responds to various challenges, such as different speeds, turning radii, or disturbances (e.g., wind and road friction).

3. Cost efficiency: Simulating the vehicle's behavior is far more cost-effective than conducting repeated real-world tests. Each test in the physical world requires fuel and equipment, as well as possibly repairing or replacing components due to wear and tear. Simulations can be run continuously without these costs, enabling thousands of iterations with minimal expenses.

4. System tuning: Through simulations, engineers can tune the vehicle's control algorithms to respond optimally to a wide variety of situations. This is

especially important for critical maneuvers like U-turns, where sharp turns at different radii and speeds, combined with disturbances (such as surface irregularities or external forces), can challenge the stability and control of the vehicle.

5.1 U-turn simulation with variations

In this section, we simulate the vehicle performing various U-turn manoeuvres as it shown in **Figure 2** under different conditions, including:

- Different radii of curvature: The radius of a U-turn can vary depending on the space available. Testing the vehicle's response to both sharp and gradual U-turns ensures that the control system can handle diverse environments, from tight urban streets to wider rural roads.

- Different speeds: Vehicle speed plays a critical role in its ability to maintain stability during turns. Simulating U-turns at different speeds helps test the vehicle's capability to stay stable and follow the intended path, especially when turning quickly, which could be necessary in emergency scenarios.

- Disturbances: External disturbances such as wind, road surface friction, or even sensor noise can affect the vehicle's performance. By treating these factors as a lumped uncertainty and incorporating them into the closed-loop control system during simulations, we demonstrated the robustness of the vehicle's control system. This ensures that the controller maintains safe operations even under less-than-ideal conditions, as evidenced by the tracking error with and without uncertainty, shown in **Figure 3**.

In summary, the validation through simulations ensures that the autonomous vehicle is thoroughly tested across various operational conditions, enabling the development of a robust and reliable system before transitioning to real-world testing. This systematic approach minimizes risks, reduces costs, and enhances safety.

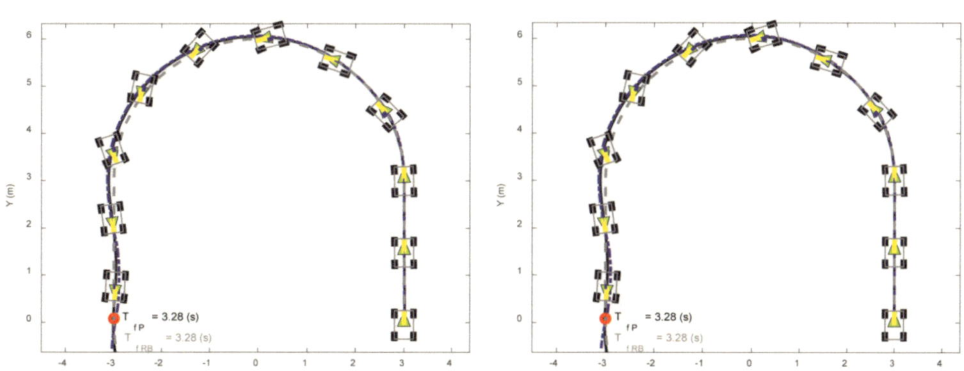

Figure 2.
Controller's performance for U-turn maneuvers of an autonomous robotic vehicle.

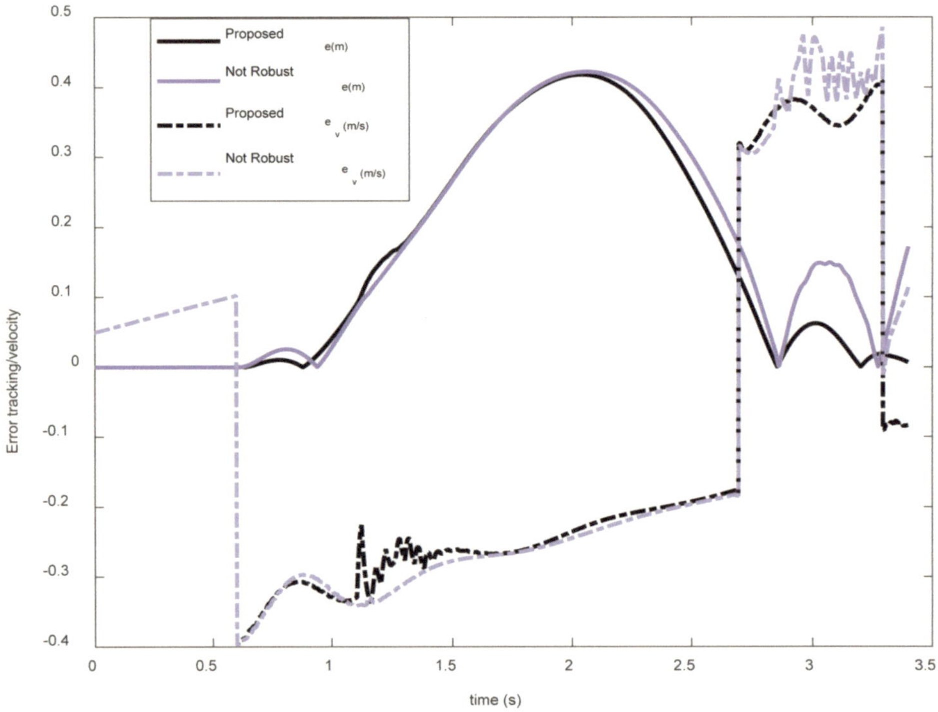

Figure 3.
Tracking error performance comparison for the controller with and without the robustness factor.

6. Conclusions

In conclusion, we have introduced a unique and innovative approach to modeling autonomous robotic vehicles, focusing on a holistic and detailed representation of their complex dynamics. This method accounts for various real-world challenges that autonomous vehicles face, such as nonlinearity, side slips, and unpredictable environmental disturbances, providing a more realistic framework for understanding vehicle behavior. At the heart of this approach is the design of a robust Nonlinear Model Predictive Controller (NMPC), which is specifically tailored to address these challenges. By leveraging advanced control techniques, the NMPC efficiently manages the vehicle's stability, ensuring smooth and accurate performance, even in the presence of uncertainties such as road friction variability, sensor noise, and external forces like wind or terrain inconsistencies.

One of the critical strengths of our approach is its focus on real-time optimization, allowing the controller to predict future vehicle states and make timely adjustments to avoid issues such as side slips, which can significantly affect vehicle safety and maneuverability. This predictive capability ensures that the autonomous vehicle can operate reliably under diverse conditions, from urban environments with sharp turns and obstacles to high-speed highway driving.

To further enhance the learning and application of this model, we have provided extensive simulations covering various test scenarios including U-turn maneuvers at different radii, speeds, and under varying levels of disturbances. These simulations not only validate the effectiveness of the proposed control system but also offer insights into the vehicle's behavior under a wide range of conditions. In addition, we offer open-source code to the community, providing a valuable resource for researchers, developers, and students interested in exploring and building upon this work. By making these tools available, we aim to foster further innovation and experimentation, helping to advance the field of autonomous vehicle control. Our contributions serve as both a foundation for understanding the complex control systems that govern autonomous vehicles and a stepping stone for future advancements in this rapidly evolving field.

Acknowledgements

I would like to extend my deepest gratitude to those who have guided and supported me throughout my academic journey. My sincere thanks go to all my academic advisors and mentors, whose encouragement and expertise have been invaluable. I am especially grateful to my MS advisor, Professor Mohammad Mahdi Fateh, and my Ph.D. advisor, Professor Jingang Yi, who inspired me to pursue an academic path through their passion and dedication to their work. Their mentorship and support have been instrumental in my growth and development. Thank you to everyone who has contributed to this work in meaningful ways.

Conflict of interest

The author declares no conflict of interest.

Thanks

I would like to express my gratitude to Professor Mahmut Reyhanoglu, academic editor, for his invaluable effort in bringing this book to publication. His expertise and dedication were essential in ensuring its quality and success. Thank you for your contributions.

Author details

Aliasghar Arab[1,2]

1 New York University, New York, NY, USA

2 North Carolina A&T State University, Greensboro, NC, USA

*Address all correspondence to: aliasghar.arab@nyu.edu; aarab@ncat.edu

References

[1] Constantinou E, Panayiotou G, Konstantinou N, Loutsiou-Ladd A, Kapardis A. Risky and aggressive driving in young adults: Personality matters. Accident; Analysis and Prevention. 2011; **43**(4):1323-1331

[2] Sparrow R, Howard M. When human beings are like drunk robots: Driverless vehicles, ethics, and the future of transport. Transportation Research Part C: Emerging Technologies. 2017;**80**:206-215

[3] Kahane CJ, Dang JN. The long-term effect of ABS in passenger cars and LTVs. Technical Report. 2009

[4] Zhou A, Wang J, Peeta S. Robust control strategy for platoon of connected and autonomous vehicles considering falsified information injected through communication links. Journal of Intelligent Transportation Systems. 2023;**27**(6):735-751

[5] Matute J, Diaz S, Karimoddini A. Sliding mode control for robust path tracking of automated vehicles in rural environments. IEEE Open Journal of Vehicular Technology. 2024;**5**:314-1325

[6] Taghavifar H, Shojaei K. Adaptive robust control algorithm for enhanced path-tracking performance of automated driving in critical scenarios. Soft Computing. 2023;**27**(13):8841-8854

[7] Liang Z, Shen M, Zhao J, Li Z, Wang Y, Ding Z. Adaptive sliding mode fault tolerant control for autonomous vehicle with unknown actuator parameters and saturated tire force based on the center of percussion. IEEE Transactions on Intelligent Transportation Systems. 2023;**24**(11): 11595-11606

[8] Feng P, Zhang J, Lu T. Feedback linearisation and disturbance observer based path following control for autonomous ground vehicle. International Journal of Vehicle Systems Modelling and Testing. 2023;**17**(3-4): 197-221

[9] Arab A, Yu K, Yu J, Yi J. Motion planning and control of autonomous aggressive vehicle maneuvers. IEEE Transactions on Automation Science and Engineering. 2023;**21**(2):1488-1500

[10] Fateh MM, Arab A. Robust control of a wheeled mobile robot by voltage control strategy. Nonlinear Dynamics. 2015;**79**:335-348

[11] Wei J, Liu YJ, Chen H, Liu L. Adaptive neural control of connected vehicular platoons with actuator faults and constraints. IEEE Transactions on Intelligent Vehicles. 2023;**8**(6):3647-3656

[12] Arab A, Fateh MM. An uncertainty compensator for robust control of wheeled mobile robots. Advanced Robotics. 2015;**29**(20):1303-1313

[13] El Hajjami L, Mellouli EM, Žuraulis V, Berrada M. A novel robust adaptive neuro-sliding mode steering controller for autonomous ground vehicles. Robotics and Autonomous Systems. 2023;**170**:104557

[14] Zhang L, Jiang Y, Chen G, Tang Y, Lu S, Gao X. Heading control of variable configuration unmanned ground vehicle using PID-type sliding mode control and steering control based on particle swarm optimization. Nonlinear Dynamics. 2023;**111**(4):3361-3378

[15] Arunprasad V, Gupta B, Karthikeyan T, Ponnusamy M. Hybrid neuro-fuzzy-genetic algorithms for optimal control of

autonomous systems. ICTACT journal on. Soft Computing. 2023;**13**(4):3015-3020

[16] Liu W, Hua M, Deng Z, Meng Z, Huang Y, Hu C, et al. A systematic survey of control techniques and applications in connected and automated vehicles. IEEE Internet of Things Journal. 2023;**10**(24):21892-21916

[17] Fateh M, Arab A. Voltage control strategy for an uncertain Mobile robot. International Journal of Intelligent Computing and Cybernetics. 2014;**7**(4):436-452

[18] Shao Y, Sun Z. Optimal speed control for a connected and autonomous electric vehicle considering battery aging and regenerative braking limits. In: Dynamic Systems and Control Conference. Vol. 59148. Park City, Utah, USA: American Society of Mechanical Engineers; 2019. p. V001T08A005

[19] Kim E, Kim J, Sunwoo M. Model predictive control strategy for smooth path tracking of autonomous vehicles with steering actuator dynamics. International Journal of Automotive Technology. 2014;**15**:1155-1164

[20] Fateh MM. On the voltage-based control of robot manipulators. International Journal of Control, Automation, and Systems. 2008;**6**(5):702-712

[21] Yu J, Arab A, Yi J, Pei X, Guo X. Hierarchical framework integrating rapidly-exploring random tree with deep reinforcement learning for autonomous vehicle. Applied Intelligence. 2023;**53**(13):16473-16486

[22] Arab A, Hadžić I, Yi J. Safe predictive control of four-wheel mobile robot with independent steering and drive. In: 2021 American Control Conference (ACC). New Orleans, LA, USA: IEEE; 2021. pp. 2962-2967

Chapter 4

Advanced Control Strategies for Space Systems: Integration of Model Predictive Control and Neural Networks

Sean Kalaycioglu and Anton de Ruiter

Abstract

This chapter presents advanced control methodologies for space systems, focusing on the integration of nonlinear model predictive control (NMPC) and neural network approaches. The chapter synthesizes novel developments in controlling coupled structural-attitude dynamics of spacecraft with flexible appendages and multiple robotic manipulators. Key innovations include the application of nonlinear autoregressive exogenous model (NARX) neural networks for adaptive state estimation, passivity-based NMPC for robust control, and piezoelectric actuator integration for precise vibration suppression. The chapter provides comprehensive coverage of mathematical modeling, control algorithm development, and practical implementation considerations. Simulation results demonstrate superior performance compared to conventional approaches, particularly in handling model uncertainties and disturbances while maintaining strict bounds on actuator saturation limits. The methodologies presented are directly applicable to emerging space applications including on-orbit servicing, assembly, and debris removal.

Keywords: model predictive control, passivity, neural networks, space robotics, multi-robot system, vibration control

1. Introduction

Model predictive control (MPC) has emerged as one of the most powerful and versatile control methodologies in modern control engineering. Its ability to handle complex, multi-variable systems while explicitly considering constraints and optimizing performance objectives has led to its widespread adoption across diverse applications. Originally developed for process industries in the 1970s, MPC has since transcended its initial domain to find applications in aerospace, robotics, and various other fields requiring sophisticated control solutions [1].

The fundamental principle of MPC lies in its predictive nature, where the controller uses a dynamic model of the system to forecast future behavior and optimize control actions over a prediction horizon. This forward-looking approach, combined with the capability to handle constraints and multiple objectives simultaneously, makes MPC particularly suitable for complex aerospace and robotic systems where precision, safety, and performance optimization are paramount.

In the realm of space and robotics applications, MPC has demonstrated remarkable versatility and effectiveness. This chapter focuses on three significant areas where MPC has made substantial contributions: free-flying autonomous multi-robot systems with passivity-based stabilization, multi-rover dual arm control, and vibration control of spacecraft antennas. These applications represent cutting-edge implementations where MPC's predictive and optimization capabilities have been instrumental in achieving superior control performance.

The integration of MPC in these domains has led to innovative solutions for challenges that traditional control methods struggle to address effectively. In multi-robot systems, MPC enables coordination while ensuring stability through passivity-based approaches. For multi-rover systems with dual arms, it provides the framework for managing complex kinematic chains while optimizing resource utilization. In spacecraft antenna control, MPC's predictive nature proves invaluable for suppressing vibrations while maintaining pointing accuracy.

The following subsections in this introduction provide comprehensive literature reviews for each of these application areas, examining the evolution of MPC implementations, key theoretical developments, and significant practical achievements. Through these reviews, we highlight how MPC has been adapted and enhanced to meet the specific challenges of each domain, while identifying common threads and potential areas for future development.

1.1 Free-flying autonomous multi-robot systems for space applications

The last two decades have witnessed a transformative shift in space operations through the advancement of free-flying autonomous multi-robot systems [1]. This evolution has been driven by the limitations and inherent risks of traditional astronaut-conducted Extravehicular Activities (EVAs), leading space agencies to pursue more efficient autonomous solutions [2]. The extensive training requirements, operational complexities, and safety considerations of human spacewalks have catalyzed the development of sophisticated robotic alternatives, particularly for operations beyond low Earth orbit.

The technological progression in this field is exemplified by several pioneering missions and initiatives. JAXA's Engineering Test Satellite ETS-VII demonstrated fundamental capabilities in autonomous capture and manipulation [3], while NASA's ARCHINAUT program [4] has advanced on-orbit manufacturing and assembly capabilities. Contemporary projects span from NASA and MAXAR Technologies' OSAM-1 mission [5] to ambitious ventures like the Gateway Moon Space Station, ESA's Mars exploration initiatives, and DLR's OrbitalHub [6].

Space debris management has emerged as a critical application driving innovation in autonomous systems. With debris mass exceeding 8000 tons in Earth orbit, the Space Foundation's 2020 white paper emphasized the urgency of debris removal operations [7]. The field has seen significant developments in capture methodologies [8], encompassing both rigid and flexible capture systems. These include innovative approaches such as harpoons, nets, and tentacles [9–11] designed to minimize collision risks and prevent additional debris generation [12]. China's Aolong-1 satellite and the Future Space Debris Removal Orbital Manipulator (FSDROM) represent practical implementations of these technologies [13].

The operational challenges in space robotics stem from multiple sources of uncertainty and system limitations. These include variations in friction coefficients, geometric properties, and dynamic parameters [14], compounded by the potential for actuator malfunctions during critical operations [15]. The complexity increases with sensor and actuator failures, requiring systems to maintain functionality with reduced degrees of freedom and limited feedback [16].

Model predictive control (MPC) has emerged as a promising framework for addressing these challenges [17]. The evolution of MPC implementations encompasses various approaches, including implicit, explicit, adaptive, and gain-scheduled variants [18–22]. While traditional linear MPC applications have shown limitations in highly nonlinear space systems [23], nonlinear model predictive control (NMPC) offers improved capabilities for handling system nonlinearities.

Recent research has focused on addressing key challenges in NMPC implementation, including feasibility, nonlinearity management, and robustness [24–26]. Various studies have explored NMPC applications in space robotics [27–32], highlighting both potential benefits and limitations. A significant concern is that NMPC alone cannot guarantee closed-loop stability, particularly crucial in space operations where system failures can have catastrophic consequences [33].

A breakthrough in addressing these limitations has emerged through the integration of passivity theory with NMPC frameworks. Building on fundamental work in passivity theory [34, 35], the researchers in Ref. [36] introduced a passivity-based NMPC scheme that provides formal stability guarantees while maintaining optimization capabilities. This approach has been further developed in recent work on passivity-based nonlinear model predictive control (PNMPC) for multi-robot space applications [37], demonstrating enhanced stability and performance in complex space operations.

The synthesis of passivity theory and NMPC represents a significant advancement in control system design for space robotics. This integrated approach addresses the dual challenges of maintaining stable operation while optimizing performance in the presence of uncertainties and constraints. As space operations continue to evolve, from debris removal to on-orbit manufacturing, these advanced control strategies will play an increasingly crucial role in enabling more complex and ambitious space missions.

The integration of mobile rovers with robotic manipulators represents a frontier in autonomous exploration and construction capabilities, particularly in planetary environments [38]. These sophisticated systems, combining mobility with manipulation capabilities, are revolutionizing planetary exploration, construction, mining operations, and military applications. The complexity of coordinating multiple rovers equipped with dual manipulators, especially when handling common loads, presents unique challenges that push the boundaries of control system design.

The evolution of rover control systems has progressed through several fundamental stages. Pioneering work in Refs. [39, 40] established crucial frameworks for managing free and contact motion control, alongside developing methodologies for generating collision-free trajectories. This foundational research was extended through investigations into nonholonomic constraints using differential wheeled rovers [41, 42], which revealed the intricate complexities of motion control under these constraints. While conventional approaches often simplified the problem by isolating kinematic control from dynamics [43], subsequent research demonstrated the feasibility of controlling mechanical systems despite their nonholonomic structure [44]. However, a significant limitation was identified in that such systems cannot achieve single-point equilibrium through smooth time-invariant feedback [45].

Early advances in multi-manipulator control were marked by Kalaycioglu [46], which demonstrated optimal force distribution techniques for dual robotic manipulators. Though limited to fixed-base systems, this research laid groundwork for later developments in mobile manipulation. The integration of model predictive control (MPC) frameworks [18, 20, 28, 47–49] brought new capabilities for performance optimization while considering system constraints and dynamics. However, implementing robust MPC posed significant challenges due to various uncertainties affecting system performance [50], although the receding horizon characteristic of standard MPC provided some inherent robustness [51].

Standard MPC implementations showed limitations in complex robotics systems [52], prompting extensive research into novel MPC methodologies [53–57] aimed at enhancing robustness and stability. The evolution of nonlinear model predictive control (NMPC) has been particularly significant, with advances in computational tools and techniques [58] dramatically improving implementation capabilities. These developments included fast gradient methods and input parameterization approaches, making NMPC increasingly practical for real-time applications.

A significant development in rover control systems was presented in Ref. [27] in the area of "Nonlinear Model Predictive Control of Rover Robotics System." This research introduced innovative approaches for managing complex nonlinear dynamics in rover systems, particularly focusing on real-time optimization and constraint handling for wheeled mobile platforms. Their work demonstrated superior performance in trajectory tracking and stability maintenance under various operating conditions, establishing new benchmarks for rover control system design.

The challenge of coordinating multiple rovers with dual manipulators handling a common load represents an underdetermined system subject to nonholonomic constraints. This configuration requires sophisticated control strategies capable of simultaneously managing both kinematic and dynamic aspects while addressing practical constraints such as wheel and joint torque saturation. The integration of predictive control methodologies with robust optimization techniques offers promising solutions for these complex scenarios.

1.2 Vibration control of spacecraft appendages

The deployment of large, lightweight flexible structures such as antennas, solar panels, and masts in orbit is essential to meet stringent mission objectives [59]. However, these structures, exhibiting low natural frequencies and damping, are susceptible to large elastic displacements due to orbital disturbances and attitude maneuvers [60]. For example, one key requirement for future Earth Observation satellites involves meeting strict pointing standards, particularly when instruments are attached to or supported by expansive flexible truss structures [61]. These spacecraft often require the capability to perform rapid and precise maneuvers without exceeding deformation limits imposed by their flexible components [62, 63]. Mechanical disturbances generated onboard by attitude control equipment and pointing mechanisms can exacerbate these challenges.

This necessitates the development of advanced vibration control strategies to ensure stability and quick vibration suppression prior to capturing data. The interaction between the dynamics of flexible and rigid bodies, often referred to as Control-Structure Interaction (CSI), adds complexity to the development of efficient spacecraft control strategies [64]. Dynamic control of satellite-based flexible structures, such as antennas and solar panels, is crucial to ensuring the stability and performance of space missions [65].

Recent years have seen an increasing focus on integrating smart materials [66], such as piezoelectric materials, Shape Memory Alloys, and magnetostrictive materials, into passive structures to actively control structural vibrations [67–69]. The field of vibration control for satellite structures has seen considerable advancements, particularly in the application of PZT actuators for both active and passive control strategies [61, 70].

The vibration control of elastic structures, particularly in aerospace applications, is a critical area of research. Various studies have explored the application of PZT actuators for active vibration suppression. For instance, Zhong et al. [71] demonstrated the effectiveness of active vibration control in space frame structures using PZT technology. PZT actuators have also been employed for deployable structures in space, as discussed by Kalaycioglu et al. [24], who explored their use for vibration control in flexible structures such as solar arrays and antennas.

Additionally, several other researchers have significantly contributed to this field. The researchers in Refs. [24, 72–74] investigated the vibration suppression of flexible manipulators through the use of adaptive structures, highlighting the potential of PZT actuators in managing vibrations in aerospace applications. Another study focused on the identification and control of vibrations using smart structures, emphasizing the role of intelligent systems in enhancing structural stability in space environments [72].

Furthermore, the study in Ref. [24] presented innovative methods for vibration suppression of space structures using smart actuators, showcasing the effectiveness of embedded PZT actuators in controlling vibrations. The integration of time-delay control strategies with fiber-optic sensors and embedded piezoelectric actuators for space structures was explored in Ref. [73], demonstrating significant improvements in vibration suppression.

The authors in Ref. [74] further investigated the minimization of vibrations in satellite antennas while managing structural configurations using adaptive structures, illustrating the practical applications and benefits of these technologies in real-world aerospace scenarios.

Several technical challenges must be addressed to develop effective real-time vibration control systems. These include the complex dynamic behavior of large, flexible structures, the nonlinearities and uncertainties associated with PZT actuators, and the limited computational resources available on spacecraft [75, 76]. Additionally, the robustness of control systems to the harsh environmental conditions of space, the disturbances related to temperature extremes, radiation, and microgravity, is a critical consideration.

Recent advancements in robust control strategies for spacecraft have shown promising results. The researchers in Ref. [77] demonstrated the effectiveness of multi-impulse robust input shaping combined with optimal control methods for flexible spacecraft, while Du and Wang [78] addressed the challenges of dynamic coupling in flexible space robots. The management of system uncertainties has been extensively studied by Yang et al. [79], who developed an integrated uncertain optimal design strategy for truss configuration and attitude-vibration control. The authors in Ref. [80] further extended this work through interval Riccati equation-based approaches for multi-objective optimal vibration control.

Intelligent adaptive control strategies, such as those proposed by Tokhi and Hossain [81], use feedforward control within an adaptive framework to optimize vibration suppression in flexible beams.

The literature on the control and dynamics of spacecraft and large deployable structures in space is extensive and diverse, as exemplified by the works of Refs. [82, 83]. The study [83] provided a notable contribution with their review of the development and challenges of large deployable structures specifically for astrophysics applications, offering insights into the progress and future prospects in this domain.

The researchers in Ref. [84] explored the minimization of vibrations in axially moving beams, providing insights into effective vibration control strategies for flexible structures. This work is foundational in understanding how vibrations in moving structures can be managed, which is crucial for the stability of spacecraft appendages. Additionally, the study in Ref. [85] examined the optimal deployment of spacecraft appendages, focusing on methods to achieve precise control during the deployment phase to ensure structural integrity and performance. Their research provided a basis for developing control algorithms that can be applied during the critical phases of spacecraft operation, such as the deployment of antennas and solar arrays. In subsequent study, Refs. [21, 86, 87] provided approximate analytical solutions for the vibrational motions of deploying appendages, offering practical approaches to predict and mitigate vibration issues during deployment.

The authors in Ref. [88] discussed the impact of uncertainties and dynamic contributions of flexible spacecraft, emphasizing the need for robust control strategies. Similarly, the study in Ref. [89] explored synergetic control approaches using thrusters and PZT devices for very flexible satellites, providing insights into advanced control mechanisms. A subsequent research in Ref. [90] focused on design of high performance attitude control system for the BIOMASS satellite, demonstrating practical control applications in ongoing missions. Lastly, the authors in Ref. [91] delved into optimal design and robust analysis for micro-vibration control in large space antennas, showcasing innovative methods for enhancing structural stability.

The challenges of rigid-flexible coupling in spacecraft have been addressed through various approaches. The study in Ref. [92] proposed effective methods for rapid attitude stabilization, while Zhang et al. [93] investigated the dynamic characteristics of vibration isolation platforms, particularly focusing on the effects of strut joints. These studies provide valuable insights into the complex interactions between rigid body motion and structural flexibility.

However, there has been a notable gap in the application of advanced control techniques such as nonlinear model predictive control (NMPC) and neural network-based control techniques such as Nonlinear AutoRegressive with eXogenous inputs (NARX) neural network for the coupled structural and attitude dynamics of flexible plate-type appendages. Advanced control strategies such as NMPC and NARX are relatively new in the context of space applications. NMPC provides robust control by predicting future system behaviors and optimizing control actions.

In addition, reinforcement learning techniques, such as Q-learning-based multi-rate optimal control, have been applied to enhance control strategies in process industries [94]. These methods offer a scalable framework for adaptive control, potentially useful for managing the dynamic behavior of flexible spacecraft structures.

NARX techniques enhance the adaptability of control systems to dynamic and uncertain environments. The combination of these techniques with PZT actuators offers a promising approach to managing the complex dynamics of flexible satellite structures. The research by Refs. [95, 96] highlights the effectiveness of NARX neural networks for vibration suppression. Their study, focusing on piezoelectric cantilever beams, demonstrated a remarkable 90.8 percent vibration rejection ratio, suggesting the potential of this technology for applications demanding high precision, such as

DOI: http://dx.doi.org/10.5772/intechopen.1010037

aerospace engineering and advanced instrumentation. Research on data-driven control and prediction for NARX systems outlined methods to predict system evolution directly from data without estimating a system plant. This approach was applied to control nonlinear systems by generating inputs for desired trajectories and tracking predicted trajectories [97].

Recent advancements have further expanded the application of NMPC and learning-based control approaches in managing complex dynamics. For instance, learning-based NMPC techniques that integrate uncertainty compensation and robust adaptation capabilities have shown significant promise in dynamic environments, as highlighted by Xie et al. [94].

Neural network-based approaches have also been extended to cooperative robotics and formation control of aerial systems, showcasing their versatility and effectiveness. For example, the research by Najafqolian et al. [37] utilized LSTM-based model predictive control for formation control of aerial robots, a method that could inspire analogous strategies for space structures. Furthermore, the authors [98] demonstrated the utility of Zeroing Neural Networks in controlling connected automated vehicles, providing insights into managing parameter-varying systems.

Similarly, the study in Ref. [99] developed a Learning-Based Explicit-MPC for aerial robots, demonstrating its potential for efficient control in dynamic scenarios. Sun et al. [100] designed a levitation predictive controller based on model predictive control aiming at the problem of levitation system constraints. These innovations underline the adaptability and precision of advanced control methods, particularly in scenarios requiring high computational efficiency and real-time optimization. These approaches provide foundational methodologies that could be adapted for space missions, where complex and uncertain conditions prevail.

In conclusion, the literature highlights the potential of advanced control strategies, such as NMPC and NARX, in combination with PZT actuators, to enhance the stability and performance of flexible satellite structures [101–103]. These advancements represent a significant step forward in the field of vibration control for space applications, offering new possibilities for the design and operation of future space missions.

The implementation of NMPC and NARX for vibration control represents a significant advancement in the field. NMPC offers robust control by predicting future system behaviors and optimizing control actions accordingly, while NARX provides adaptive capabilities that enhance the system's response to unknown dynamic and uncertain environments. These control strategies are applied to the coupled dynamics model, enabling the precise control of the flexible antennas equipped with PZT sensors and actuators. The employment of PZT actuators for vibration suppression is particularly advantageous due to their high actuation precision, lightweight nature, energy efficiency, and dual functionality as both sensors and actuators. These attributes make them ideal for the demanding conditions of space applications.

The remainder of this chapter is structured as follows. Section 2 presents the mathematical model that forms the foundation of the study, detailing the governing equations, underlying assumptions, and key formulations necessary for analysis. Section 3 provides the simulation results, demonstrating the application of the proposed model and offering a comprehensive evaluation of its performance under various conditions. The results are analyzed in depth to assess their implications and validate the theoretical framework. Finally, Section 4 concludes the chapter by summarizing the key findings, discussing their broader significance, and outlining potential directions for future research.

2. Mathematical modeling

This section presents the mathematical models essential for analyzing and controlling space-based robotic and structural systems. The first part (Section 2.1) focuses on free-flying autonomous multi-robot systems for space applications. It introduces the dynamic modeling of a chaser spacecraft equipped with two redundant robotic manipulators, describing the system's kinematics, equations of motion, and control strategies. The formulation incorporates passivity-based nonlinear model predictive control (NMPC) to ensure stability and optimize motion planning in microgravity environments.

The second part (Section 2.2) addresses the vibration control of spacecraft appendages using model predictive control (MPC). This subsection develops a coupled structural and attitude dynamics model, incorporating flexible appendages equipped with piezoelectric (PZT) actuators for active vibration suppression. Advanced control methodologies, including NMPC and an AI-driven Nonlinear AutoRegressive with Exogenous Inputs (NARX) model, are introduced to enhance stability and performance under operational uncertainties.

These mathematical formulations provide the theoretical foundation for the subsequent simulation and analysis presented in Section 3, where their effectiveness is demonstrated under realistic operational conditions.

2.1 Free-flying autonomous multi-robot systems for space applications

The studied system consists of a chaser spacecraft, a target payload, and two redundant robotic manipulators with n degrees of freedom. An illustrative depiction of such a system, where the chaser spacecraft is equipped with two robotic arms and a rigid payload, is presented in **Figure 1**.

The equations governing the motion of this system are structured as follows:

$$\mathbf{M}\ddot{\mathbf{q}} + \mathbf{C}(\mathbf{q}, \dot{\mathbf{q}})\dot{\mathbf{q}} = \tau, \tag{1}$$

where:

- $\mathbf{q}$ represents the system's generalized coordinates, including linear/angular positions of the chaser and joint angles of the robotic arms.

Figure 1.
Conceptual representation of a multi-robot spacecraft system.

- $\mathbf{M}$ is the mass/inertia matrix, which is positive definite.
- $\mathbf{C}(\mathbf{q}, \dot{\mathbf{q}})$ accounts for the Coriolis and centrifugal effects.
- τ denotes the external control forces and moments acting on the spacecraft and joint torques applied to the robotic arms.

To express the dynamics in the Cartesian space for the robotic manipulators and in Euler angles for the spacecraft, the system Jacobian $\mathbf{J}$ is introduced, facilitating the transformation from joint space to Cartesian space and from body-fixed angular rates to Euler rates:

$$\dot{\mathbf{x}} = \mathbf{J}\dot{\mathbf{q}}, \tag{2}$$

where:

- $\mathbf{x}$ includes the Cartesian positions of the spacecraft center of mass and the end-effectors of the robotic arms.
- The rotational transformation follows a 1–3-2 Euler sequence, defined by angles ϕ, ψ, θ.
- The pseudo-inverse Jacobian is utilized to derive equivalent representations of mass/inertia matrix, Coriolis matrix, and force vectors in the Cartesian domain.

The system dynamics in the transformed coordinates can be represented as:

$$\mathbf{M}_c\ddot{\mathbf{x}} + \mathbf{C}_c\dot{\mathbf{x}} = \mathbf{f}_c, \tag{3}$$

where $\mathbf{M}_c$, $\mathbf{C}_c$, and $\mathbf{f}_c$ are the equivalent matrices and force vectors in the Cartesian representation. The output $\mathbf{y}$ can be written as:

$$\mathbf{y} = \begin{bmatrix} \mathbf{x} \\ \dot{\mathbf{x}} \end{bmatrix}. \tag{4}$$

To ensure clarity in the formulation of the multi-robot spacecraft dynamics, **Table 1** provides a summary of all key variables used in the mathematical modeling. This includes generalized coordinates, mass and inertia properties, Coriolis effects, external control forces, and their representations in both joint and Cartesian spaces.

2.1.1 Passivity and passivity-based NMPC

A system is defined as passive if there exists a positive semi-definite storage function $V(x)$ satisfying [37]:

$$\dot{V} \leq \mathbf{u}^T\mathbf{y}, \tag{5}$$

where $\mathbf{u}$ and $\mathbf{y}$ denote the input and output signals, respectively.

Symbol	Description
$\mathbf{q}$	Generalized coordinates, including linear/angular positions of the chaser and joint angles of the robotic arms.
$\mathbf{M}$	Positive definite mass/inertia matrix.
$\mathbf{C}(\mathbf{q},\dot{\mathbf{q}})$	Coriolis and centrifugal effects matrix.
τ	External control forces and moments on the spacecraft, including joint torques of robotic arms.
$\mathbf{x}$	Cartesian positions of the spacecraft center of mass and the end-effectors of the robotic arms.
ϕ,ψ,θ	Rotational transformation angles following a 1–3-2 Euler sequence.
$\mathbf{J}^{\dagger}$	Pseudo-inverse Jacobian used to derive equivalent representations of mass/inertia matrix, Coriolis matrix, and force vectors in the Cartesian domain.
$\mathbf{M}_c$	Equivalent mass/inertia matrix in the Cartesian representation.
$\mathbf{C}_c$	Equivalent Coriolis and centrifugal effects matrix in the Cartesian representation.
$\mathbf{f}_c$	Equivalent force vector in the Cartesian representation.
$\mathbf{y}$	Output vector containing the system states, defined as $\mathbf{y}=\begin{bmatrix}\mathbf{x}\\ \dot{\mathbf{x}}\end{bmatrix}$.

Table 1.
Summary of variables used in multi-robot dynamics equations.

A more refined definition, termed Input Feed-forward Output Feedback Passive (IF-OFP), involves additional constraints:

$$\dot{V}\leq\mathbf{u}^T\mathbf{y}-\rho\|\mathbf{y}\|^2-\nu\|\mathbf{u}\|^2. \tag{6}$$

The positive indices ρ and ν characterize the degree of passivity.

The nonlinear model predictive control (NMPC) framework incorporates passivity concepts to ensure stability. The optimization problem is formulated as:

$$\min V\left(\mathbf{y}_{N_p}\right)+\frac{1}{2}\int_0^{T_p}\left[(\mathbf{y}(t)-\mathbf{y}_r(t))^T\mathbf{Q}(\mathbf{y}(t)-\mathbf{y}_r(t))+\mathbf{u}^T(t)\mathbf{R}\mathbf{u}(t)\right]dt \tag{7}$$

subject to:

$$\mathbf{x}_{k+1}=f(\mathbf{x}_k,\mathbf{u}_k), \tag{8}$$

where T_p is the time value corresponding to the prediction horizon (N_p), and $\mathbf{Q},\mathbf{R}$ are positive-definite weighting matrices. $\mathbf{y}(t)$, $\mathbf{y}_r(t)$, and $\mathbf{y}_m(t)$ are the plant output, the reference trajectory, and the model output, respectively. To mitigate model uncertainties, a bias correction mechanism is implemented as shown in **Figure 2**.

The bias-adjusted model predictions are computed as:

$$\hat{\mathbf{y}}_j=\mathbf{y}_j+\mathbf{d}_j,\quad j=1,\ldots,N_p, \tag{9}$$

where $\mathbf{d}_j$ represents the bias correction term.

To ensure closed-loop stability, a terminal cost function is incorporated:

$$V_f\left(\mathbf{x}_{N_p}\right)\geq\lambda\|\mathbf{x}_{N_p}\|^2, \tag{10}$$

which enforces the passivity constraints and guarantees stability.

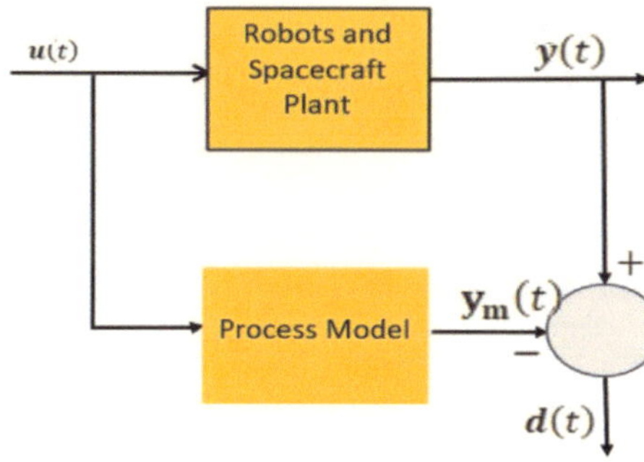

Figure 2.
Bias computation using an independent model.

To further refine stability analysis, we introduce the tracking error vector:

$$\mathbf{e} = \mathbf{x} - \mathbf{x}_{\mathrm{r}}, \tag{11}$$

and define the storage function:

$$V(\mathbf{y}_t) = \frac{1}{2}\dot{\mathbf{x}}^T \mathbf{M}_c \dot{\mathbf{x}} + \frac{1}{2}\mathbf{e}^T \mathbf{K}\mathbf{e} \tag{12}$$

where $\mathbf{K}$ is a positive definite matrix. The derivative of $V(\mathbf{y}_t)$ is then computed using:

$$\dot{V}(\mathbf{y}_t) = [\mathbf{f}_c + \mathbf{K}\mathbf{e}]^T \dot{\mathbf{x}}. \tag{13}$$

The following terminal value of the predicted control input at $t = T_P$ satisfies the passivity-based inequality constraints:

$$\mathbf{u}(t) = \mathbf{u}(T_p) = [\mathbf{f}_c + \mathbf{K}\mathbf{e}] \tag{14}$$

Therefore,

$$\dot{V}(T_p) = [\mathbf{f}_c + \mathbf{K}\mathbf{e}]^T \dot{\mathbf{x}} \leq \mathbf{u}^T(t)\dot{\mathbf{x}} \tag{15}$$

This ensures closed-loop stability while incorporating model parameter uncertainties through a bias correction term $\mathbf{K}\mathbf{e}$.

The proposed passivity-based NMPC (PNMPC) scheme integrates state constraints, terminal cost functions, and passivity properties, providing an innovative approach to controlling nonlinear space multi-robot systems.

This methodology offers a robust alternative to the control Lyapunov function by ensuring stability and performance while leveraging the benefits of NMPC optimization.

The framework demonstrates the feasibility of passivity-based NMPC for multi-robot spacecraft applications, merging nonlinear control techniques with optimality constraints to ensure enhanced stability and performance.

To summarize the key variables and parameters used in the passivity-based NMPC formulation, **Table 2** provides a comprehensive list, including the control input and

Symbol/parameter	Description
N_p	Prediction horizon.
$\mathbf{Q}, \mathbf{R}$	Cost function weighting matrices.
V_f	Terminal cost function.
ρ, ν	Positive indices characterizing the degree of passivity.
T_p	Time value corresponding to the prediction horizon (N_p).
$V(x)$	Storage function used in passivity analysis.
$\mathbf{u}$	Control input vector.
$\mathbf{y}$	Output vector containing system states.
$V_f\left(\mathbf{x}_{N_p}\right)$	Terminal cost function enforcing stability constraints.
$\mathbf{d}_j$	Bias correction term used to mitigate model uncertainties.
$\mathbf{e}$	Tracking error vector, defined as $\mathbf{x} - \mathbf{x}_r$.
$\mathbf{K}$	Positive definite matrix in the storage function.
$\mathbf{M}_c$	Equivalent mass/inertia matrix in the Cartesian representation.
$\mathbf{C}_c$	Equivalent Coriolis and centrifugal effects matrix in the Cartesian representation.
$\mathbf{f}_c$	Equivalent force vector in the Cartesian representation.
$\hat{\mathbf{y}}_j$	Bias-adjusted model predictions.
$\mathbf{y}_{N_p}$	System output at the final prediction step.
$\mathbf{u}(T_p)$	Terminal value of the predicted control input.

Table 2.
Summary of variables used in passivity-based NMPC.

output variables, stability parameters, cost function weights, and tracking error components.

2.2 Vibration control of spacecraft appendages using MPC

The system considered in this study consists of a rigid-body spacecraft equipped with multiple long, flexible plate-like appendages, each embedded with piezoelectric (PZT) actuators for active vibration control.

A local coordinate system (x_i, y_i, z_i) is assigned to each appendage. The x_i-axis aligns with the neutral axis of the appendage, and the coordinate system maintains a fixed inclination with respect to the body-fixed frame (X, Y, Z). The attachment point of each appendage to the spacecraft is offset from the center of mass by $\overline{d}_{1i}, \overline{d}_{2i}, \overline{d}_{3i}$, corresponding to displacements along the X, Y, and Z-axes, respectively. **Figure 3** illustrates such a system.

Each appendage is modeled as a flexible plate with uniform cross-sectional area A_i, a moment of inertia I_i about its neutral axis, and a mass distribution of m_i per unit length along the x-direction. The material properties are characterized by the elasticity modulus E_i and Poisson's ratio ν_i. The geometric parameters include the length a_{io}, width b_{io}, and thickness h_{io}, while the embedded piezoelectric actuators have

DOI: http://dx.doi.org/10.5772/intechopen.1010037

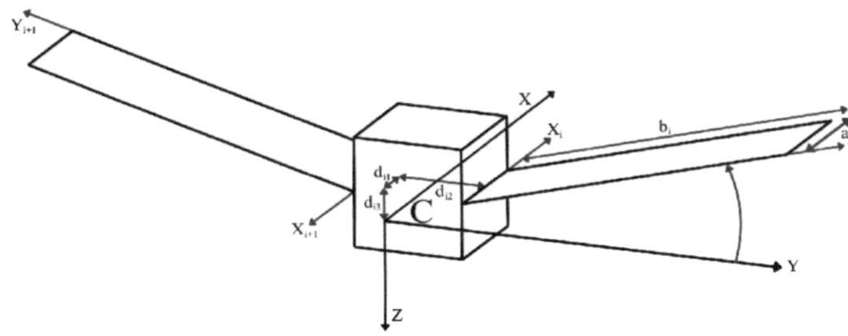

Figure 3.
Schematic representation of a rigid spacecraft with flexible appendage.

dimensions a_{ij}, b_{ij}, h_{ij} along their respective axes. Here, $i = 1, \dots, n$ denotes the number of appendages, and $j = 1, \dots, m$ represents the total number of actuators.

A schematic representation of a plate-type appendage equipped with PZT actuators is shown in **Figure 4**.

2.2.1 Coupled structural and attitude dynamics

The spacecraft's attitude and structural vibrations are interconnected through the system's energy equations. The total kinetic energy of the system accounts for both the rigid-body dynamics and the flexible appendages, given by:

$$T_{total} = \frac{1}{2}\omega^T[I_c + I_r + I_e]\omega + \omega^T \mathbf{K}_e + T_e + T_c + T_{orb}. \tag{16}$$

The symbols and parameters used in the formulation of the kinetic energy expression are summarized in **Table 3**, which provides definitions for inertia matrices, kinetic energy terms, and angular velocity components.

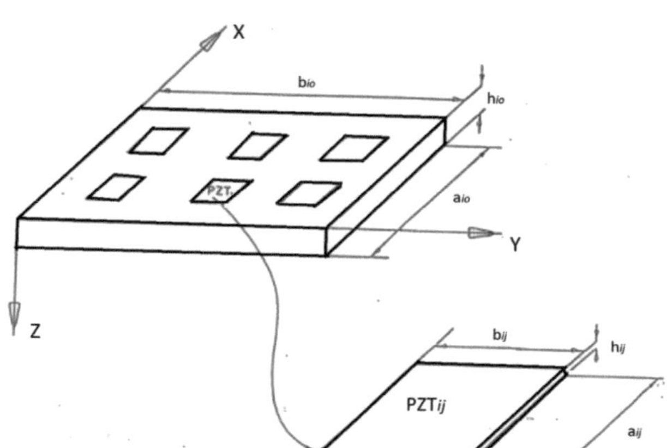

Figure 4.
Schematic representation of a flexible appendage with embedded piezoelectric actuators.

Symbol	Description
I_c	Inertia matrix of the rigid spacecraft
I_r	Inertia matrix of the appendages in their undeformed (rigid body) state
I_e	Contribution to the inertia matrix due to elastic deformations
T_e	Kinetic energy associated with transverse displacements of the appendages
$\mathbf{K}_e$	Vector dependent on elastic deformations
T_{orb}	Kinetic energy resulting from the orbital motion of the center of mass
T_c	Kinetic energy of the central body
ω	Angular velocity components of the spacecraft in the body-fixed frame

Table 3.
Description of symbols used in the kinetic energy expression.

The elastic deformation $u_i(x_i, y_i, t)$ of the flexible appendages is represented using a modal expansion:

$$u_i(x_i, y_i, t) = \sum_{s=1}^{n_s}\sum_{r=1}^{n_r} W_{irs}(t)\phi_r(x_i)\psi_s(y_i) = \Phi_i^T W \Psi_i. \tag{17}$$

where $\phi_r(x_i)$ and $\psi_s(y_i)$ define the mode shapes satisfying the boundary conditions, and $W_{irs}(t)$ are the modal displacement coefficients.

The potential energy of the system, incorporating the bending and shear deformations, is expressed as:

$$U_e = \sum_{i=1}^{n} \frac{1}{2} D_i \int_0^{b_i}\int_0^{a_i} \left[\left(\frac{\partial^2 u_i}{\partial x_i^2}\right)^2 + \left(\frac{\partial^2 u_i}{\partial y_i^2}\right)^2 + 2(1-\nu)\left(\frac{\partial^2 u_i}{\partial x_i \partial y_i}\right)^2 \right] dx_i dy_i. \tag{18}$$

where D_i is the flexural rigidity given by:

$$D_i = \frac{E_{io} h_{io}^3}{12(1-\nu^2)}. \tag{19}$$

Using Lagrangian mechanics, the governing equations for the coupled structural and attitude motion are derived and presented below:

$$\ddot{\vec{W}}_i + D_{wi}\dot{\vec{W}}_i + S_{1i}\vec{W}_i = \vec{S}_{2i} \tag{20}$$

which governs the vibrations of the i-th appendage. D_{wi} is the damping matrix, assumed to have a diagonal form. The matrix S_{1i} and the vector $\vec{S}_{2i}$ are given by:

$$\begin{aligned} S_{1i} = &\Big[K_i + \frac{1}{b_{io}}\big(\omega_{zi}^2 d_{i2} + \omega_{xi}^2 d_{i2} + \dot{\omega}_{xi} d_{i3} \\ &- \dot{\omega}_{zi} d_{i1} - \omega_{yi}\omega_{zi} d_{i3} - \omega_{xi}\omega_{zi} d_{i1}\big) C_1\Big] \\ &+ \Big[-\frac{a_{io}}{b_{io}}\big(\omega_{xi}\omega_{yi} + \dot{\omega}_{zi}\big) C_2 \\ &+ \big(\omega_{xi}^2 - \omega_{yi}^2\big)1 + \big(\omega_{xi}^2 + \omega_{zi}^2\big) C_7\Big] \end{aligned} \tag{21}$$

$$\begin{aligned}\vec{S}_{2i} = &\Big[\vec{F}_{pi} + \vec{F}_i + \Big(\omega_{yi}^2 d_{i3} + \omega_{xi}^2 d_{i3} - \dot{\omega}_{xi} d_{i2} \\ &+\dot{\omega}_{yi} d_{i1} - \omega_{yi}\omega_{zi} d_{i2} - \omega_{xi}\omega_{zi} d_{i1}\Big)\vec{C}_3\Big] \\ &-\Big[\big(\dot{\omega}_{xi} + \omega_{yi}\omega_{zi}\big) b_{io}\vec{C}_4 \\ &+\big(\dot{\omega}_{yi} - \omega_{xi}\omega_{zi}\big) a_{io}\vec{C}_5\Big]\end{aligned} \tag{22}$$

The constant matrices C_1, C_2, C_7 and the vectors $\vec{C}_3$, $\vec{C}_4$, $\vec{C}_5$ and $\vec{F}_i$ are provided in [104]. $\vec{F}_{pi}$ is the PZT-created control force.

The rotational dynamics of the system is described by the following matrix equation:

$$(I_r + I_e + I_c)\dot{\vec{\omega}} + [I_e + \Omega(I_r + I_e + I_c)]\vec{\omega} + \Omega\vec{K}_e + \vec{K}_e = \vec{\tau} \tag{23}$$

where Ω represents the skew-symmetric matrix associated with the angular velocity components:

$$\Omega = \begin{bmatrix} 0 & -\omega_z & \omega_y \\ \omega_z & 0 & -\omega_x \\ -\omega_y & \omega_x & 0 \end{bmatrix} \tag{24}$$

The term $\vec{\tau}$ denotes the total torque acting on the system, encompassing both control and external disturbances and can be expressed as:

$$\vec{\tau} = \vec{\tau}_c + \vec{\tau}_{gg} + \vec{\tau}_{sol} + \vec{\tau}_{aer} + \vec{\tau}_{mag} \tag{25}$$

In this equation, $\vec{\tau}_c$ represents the control torque applied to the satellite, while $\vec{\tau}_{sol}$, $\vec{\tau}_{gg}$, $\vec{\tau}_{mag}$, and $\vec{\tau}_{aer}$ correspond to torques induced by solar radiation pressure, gravity gradient effects, Earth's magnetic field interactions, and aerodynamic forces, respectively.

The key symbols and parameters used in the formulation of the coupled structural and attitude dynamics equations are summarized in **Table 4**. This table provides definitions for inertia properties, kinetic energy terms, angular velocity components, and elastic deformation variables, ensuring clarity in the mathematical modeling of the system.

2.2.2 NMPC control model

Two sophisticated control methodologies are formulated and implemented to regulate the coupled vibrational and attitude dynamics of a satellite with PZT-mounted plate-type appendages. The first approach utilizes nonlinear model predictive control (NMPC), which predicts future system behavior over a finite horizon and determines optimal control inputs accordingly. The second method employs an AI-driven Nonlinear AutoRegressive with eXogenous inputs (NARX) model, designed to function without an explicit mathematical model of the system dynamics.

This section introduces the NMPC framework for controlling the structural and vibrational behavior of PZT-mounted satellite appendages. The classical model predictive control (MPC) formulation for this system is given in Ref. [37]:

Symbol	Description
T_{total}	Total kinetic energy of the spacecraft and its appendages.
I_c	Inertia matrix of the rigid spacecraft.
I_r	Inertia matrix of the appendages in their undeformed (rigid body) state.
I_e	Contribution to the inertia matrix due to elastic deformations.
T_e	Kinetic energy associated with transverse displacements of the appendages.
$\mathbf{K}_e$	Vector dependent on elastic deformations.
T_{orb}	Kinetic energy resulting from the orbital motion of the center of mass.
T_c	Kinetic energy of the central body.
ω	Angular velocity components of the spacecraft in the body-fixed frame.
$u_i(x_i, y_i, t)$	Elastic deformation of the i-th appendage.
$\phi_r(x_i), \psi_s(y_i)$	Mode shapes satisfying boundary conditions.
$W_{irs}(t)$	Modal displacement coefficients.
D_i	Flexural rigidity of the appendages.
E_{io}, h_{io}, ν	Elasticity modulus, thickness, and Poisson's ratio of the appendages.
D_{wi}	Damping matrix of the i-th appendage (assumed diagonal).
$S_{1i}, \vec{S}_{2i}$	Structural matrices governing the vibrations of the i-th appendage.
Ω	Skew-symmetric matrix of angular velocity components.
$\vec{\tau}$	Total torque acting on the system.
$\vec{\tau}_c$	Control torque applied to the satellite.
$\vec{\tau}_{sol}, \vec{\tau}_{gg}, \vec{\tau}_{mag}, \vec{\tau}_{aer}$	Torques induced by solar radiation pressure, gravity gradient, Earth's magnetic field, and aerodynamic forces, respectively.

Table 4.
Summary of variables used in coupled structural and attitude dynamics.

$$\min \int_0^{T_p} \left(\left(\vec{y}(t) - \vec{y}_r(t) \right)^T K_y \left(\vec{y}(t) - \vec{y}_r(t) \right) + \vec{S}^T(t) K_s \vec{S}(t) \right) dt \tag{26}$$

$$\dot{\vec{y}} = \vec{g}\left(\vec{y}\right) + L\vec{S}, \quad \text{and} \quad \vec{z} = \vec{g}_z\left(\vec{y}\right) + H\vec{S} \tag{27}$$

$$\vec{y}(0) = \vec{y}(t_0) \tag{28}$$

$$\vec{S}_{\min} < \Lambda\vec{S} < \vec{S}_{\max} \tag{29}$$

where $\vec{y} = \left[\vec{y}_p; \dot{\vec{y}}_p\right]$, and $\vec{y}_p^T = \left[\vec{W}_1, \vec{W}_2, \dots, \vec{W}_n, \varphi, \psi, \theta\right]$, with n denoting the number of appendages. The reference state vector $\vec{y}_r$ defines the target values, where the generalized vibrational states $\vec{W}_i$ are zero. The predictive control horizon is denoted by T_p, and the weighting matrices K_y and K_s are positive definite.

The nonlinear system is approximated using a quasi-linear parameter-varying framework. The discretization process assumes that the continuous-time model behaves in a quasi-linear manner over each discrete time step:

$$\vec{y}(k+1) = \hat{A}(\hat{g}(k))\vec{y}(k) + \hat{B}(\hat{g}(k))\vec{S}(k) \tag{30}$$

$$\vec{z}(k) = \hat{C}(\hat{g}_z(k))\vec{y}(k) + \hat{D}(\hat{g}(k))\vec{S}(k) \tag{31}$$

$$\hat{g}(k) = f_g\left(\vec{y}(k)\right) \tag{32}$$

$$\vec{S}_{\min} \leq \Lambda\vec{S}(k) \leq \vec{S}_{\max} \tag{33}$$

At each discrete step k, the system's output $\vec{z}(k)$ is determined based on real-time measurements.

The NMPC algorithm updates the system states $\vec{y}(k)$ and control inputs $\vec{S}(k)$ at each sampling interval by minimizing the following cost function:

$$\begin{aligned} C_n = \frac{1}{2}\sum_{j=1}^{N_P}\Big(&\left(\vec{y}(k+j) - \vec{y}_r(k+j)\right)^T \\ &\times K_y\left(\vec{y}(k+j) - \vec{y}_r(k+j)\right) \\ &+ \vec{S}(k+j-1)^T K_s \vec{S}(k+j-1)\Big) \end{aligned} \tag{34}$$

subject to:

$$\vec{y}(k+j+1) = \hat{A}(\hat{g}(k+j))\vec{y}(k+j) + \hat{B}(\hat{g}(k+j))\vec{S}(k+j) \tag{35}$$

$$\vec{z}(k+j) = \hat{C}(\hat{g}(k+j))\vec{y}(k+j) + \hat{D}(\hat{g}(k+j))\vec{S}(k+j) \tag{36}$$

$$\vec{S}_{\min} \leq \Lambda\vec{S}(k+j) \leq \vec{S}_{\max} \tag{37}$$

Since active constraints affect the optimization process, the solution must be derived using the *Karush-Kuhn-Tucker (KKT) conditions*, ensuring the optimal control solution meets imposed constraints.

To solve this constrained optimization problem, the MATLAB Quadratic Programming (QP) solver is utilized, which inherently satisfies the conditions from Eq. (37). The QP solver minimizes the following quadratic cost function:

$$C_n = \frac{1}{2}\sum_{j=1}^{N_P}\left[\vec{S}(k+j)^T \hat{P}\vec{S}(k+j) + 2\left(\vec{g}\right)^T \vec{S}(k+j)\right] \tag{38}$$

where:

$$\hat{P} = \left(\hat{B}^T K_y \hat{B} + K_s\right) \tag{39}$$

$$\left(\vec{g}\right)^T = \left(\vec{y}(k+j)^T \hat{A}^T K_y \hat{B} - \vec{y}_r(k+j)^T K_s \hat{B}\right) \tag{40}$$

Symbol	Description
$\vec{y}$	System state vector, including vibrational states and attitude variables.
$\vec{y}_r$	Reference state vector, defining target system states.
$\vec{W}_i$	Generalized vibrational states of the appendages.
φ, ψ, θ	Spacecraft attitude angles.
n	Number of appendages in the system.
T_p	Prediction horizon.
K_y, K_s	Positive definite weighting matrices for the cost function.
$\vec{S}$	Control input vector.
$\vec{y}(k+j)$	System state at prediction step $k+j$.
$\hat{A}, \hat{B}, \hat{C}, \hat{D}$	Matrices approximating system dynamics in discrete time.
$\hat{g}(k)$	Quasi-linear parameter-varying system function.
$\vec{z}(k)$	System output at discrete step k.
Λ	Constraint matrix for control inputs.
$\vec{S}_{\min}, \vec{S}_{\max}$	Minimum and maximum control input constraints.
C_n	Cost function to be minimized in NMPC.
$\hat{P}$	Matrix defining the quadratic cost in the optimization problem.
$\vec{g}$	Gradient term used in quadratic programming.

Table 5.
Summary of variables used in NMPC control model.

By implementing this optimization framework, the NMPC strategy efficiently governs the satellite's structural and vibrational dynamics, ensuring stability while satisfying input constraints.

The key symbols and parameters used in the NMPC control model are summarized in **Table 5**, providing definitions for system states, control inputs, prediction horizons, and optimization parameters.

2.2.3 NARX model for adaptive vibration control

The *Nonlinear AutoRegressive with eXogenous inputs (NARX)* model [95] is a data-driven AI approach used for system identification and control, particularly effective for highly nonlinear and complex systems where conventional mathematical models may fall short [95, 96]. Unlike NMPC, the NARX model eliminates the need for an explicit analytical representation of system dynamics, making it a flexible and adaptive solution.

For the suppression of vibrations in *PZT-mounted plate-type satellite appendages*, the NARX model is leveraged to estimate system states and predict optimal control actions. The applied control inputs $\vec{S}(t)$, which include both PZT actuation forces and satellite control torques, are derived from the predicted system outputs, ensuring effective vibration attenuation.

DOI: http://dx.doi.org/10.5772/intechopen.1010037

NARX neural networks are particularly advantageous in scenarios where system behavior exhibits strong nonlinearity and requires continuous adaptation. By employing inverse dynamics learning, the NARX-based controller directly models the inverse system response, allowing it to establish a near-ideal transfer function between the target and actual outputs.

A key strength of the NARX-based approach lies in its *adaptive learning capability*. The model dynamically updates itself in response to real-time data, making adjustments to reflect variations in system behavior and external disturbances. This is particularly beneficial in space applications, where unpredictable operational conditions are commonplace [95, 96].

The control framework follows a structured adaptation process:

1. Continuously acquire real-time telemetry from the spacecraft's onboard sensors.

2. Periodically update the NARX model by retraining it with newly acquired data to reflect any changes in system dynamics.

3. Compute updated control actions $\vec{S}(t)$ using the refined model to maintain precise system regulation.

To further enhance adaptability, the *NARX model is extended with a neural network discriminator (NND)* [96]. The NND functions as an online learning module, continuously refining the weight connections of the NARX model by distinguishing between observed and predicted input-output relationships. This real-time learning mechanism significantly improves the controller's resilience against unexpected disturbances and time-dependent parameter variations, overcoming one of the major limitations of standard direct inverse control.

By integrating the NND into the adaptive control framework, the NARX-based system achieves high accuracy and robustness even in rapidly evolving environments. This makes it an ideal candidate for spacecraft and robotic systems operating in complex and uncertain conditions.

Key benefits of the NARX-based control strategy are:

- *Capturing nonlinear dynamics*: The NARX model inherently represents complex nonlinear input-output relationships, making it highly effective for systems with intricate dynamics.

- *Continuous adaptation*: The incorporation of an NND enables real-time learning and adjustment, allowing the model to evolve alongside changing system conditions.

- *Enhanced robustness*: The adaptive learning structure strengthens the controller's ability to counteract perturbations, model uncertainties, and parameter drift.

The Nonlinear AutoRegressive with eXogenous inputs (NARX) model is employed to enhance vibration control by predicting future system behavior using past inputs and outputs. The general NARX formulation is:

$$y(t+1) = h\big(y(t), y(t-1), \ldots, y(t-d_y), S(t), S(t-1), \ldots, S(t-d_s)\big), \tag{41}$$

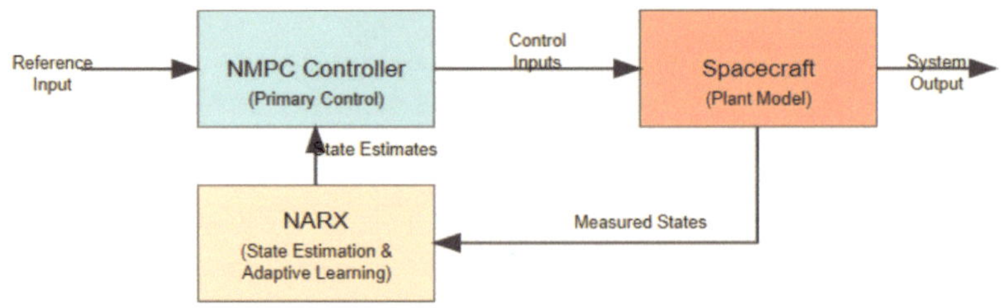

Figure 5.
Integrated NMPC-NARX framework for spacecraft vibration control.

where $y(t)$ represents the system output at time t, and $S(t)$ denotes the control input. The parameters d_y and d_s define the maximum lags in the output and input, respectively.

The mathematical formulation of the NARX-based control model is given by:

$$\tilde{y}(t+1) = \tilde{A}\tilde{y}(t) + \tilde{B}\tilde{S}(t), \tag{42}$$

$$\tilde{z}(t) = \tilde{C}\tilde{y}(t) + \tilde{D}\tilde{S}(t), \tag{43}$$

where $\tilde{A}, \tilde{B}, \tilde{C}, \tilde{D}$ are model parameters estimated using historical data.

Implementation strategy: The NARX-based control implementation follows these steps:

- Collect historical system data, including output $\tilde{y}(t)$ and input $\tilde{S}(t)$.
- Train the NARX model to capture system dynamics.
- Utilize the trained NARX model to predict future states $\tilde{y}(t+1)$.
- Compute the required control input $\tilde{S}(t)$ to achieve the desired system performance.

Finally, a hybrid approach integrating NMPC and NARX offers a robust control framework by leveraging NMPC's optimization-based control with NARX's ability to handle model uncertainties and external disturbances, as illustrated in **Figure 5**. This approach significantly improves performance in suppressing vibrations in the spacecraft appendages, ensuring stability and precise attitude control.

3. Simulation results

To thoroughly analyze the performance of NMPC, NARX-based control, and the hybrid approach, multiple disturbance conditions were introduced. These scenarios included stochastic vibrations to mimic small-scale perturbations and assess the control system's effectiveness in damping minor oscillations, sudden impulses to examine stability and reaction time under abrupt force applications, and sustained external loads to evaluate long-term disturbance rejection.

To ensure realistic conditions, noise was explicitly incorporated into the simulation by applying Gaussian-distributed variations to the vibration and impact sensor readings, with a mean of zero and a standard deviation of 0.05.

The disturbance models were carefully defined as follows:

- *Random vibrations*: Simulated using a white noise process characterized by a standard deviation of 0.1 and a bandwidth of 5 Hz.

- *Sudden impacts*: Modeled as impulse forces varying in magnitude from 5 to 20 N to evaluate the control system's rapid response capabilities.

- *Continuous external forces*: Applied as sinusoidal disturbances with frequencies between 0.5 and 2 Hz and amplitudes ranging from 0.2 to 1 N, enabling an assessment of long-term stability control.

The NARX neural network effectively captured the underlying system behavior and generated predictive estimates based on historical input-output relationships. The effectiveness of the control strategies was quantified using key performance indicators such as vibration amplitude attenuation, response speed, and overall system stability.

The specific characteristics of the PZT sensors and actuators, spacecraft structure, and flexible plate-type appendages utilized in the simulations are provided in **Table 6**.

The effectiveness and the model's predictive performance of the NARX neural network was assessed using multiple metrics and visual representations. The following figures provide a detailed examination of the results.

The training evolution of the neural network is depicted in **Figure 6**. Training was concluded after 14 epochs, as further iterations did not yield noticeable improvements in validation accuracy. This suggests that the model effectively captured the system's underlying behavior.

Parameter	Values	Unit
Plate dimensions (a_{io}, b_{io}, h_{io})	$(0.3, 1, 2 \times 10^{-3})$	m
PZT dimensions (a_{ij}, b_{ij}, h_{ij})	$(0.1, 0.1, 0.5 \times 10^{-3})$	m
Young's modulus of elasticity E_{io}, E_{ij}	6.9×10^{10}	Pa
Locations of PZTs (x_{ij}, y_{ij})	$[(0.1, 0.1), (0.5, 0.1), (0.9, 0.1)]$	m
Poisson's ratio	0.33	—
Density of the plate ρ	2.7×10^3	kg/m^3
PZT electric coefficient d_{31}	-1.75×10^{-10}	V^{-1}
Moments of inertia of the satellite	$\begin{bmatrix} 1.2 & -0.05 & -0.03 \\ -0.05 & 3.02 & -0.02 \\ -0.03 & -0.02 & 8.02 \end{bmatrix}$	kg-m^2

[a]*All values are used for simulation purposes.*[b]*PZT placement locations are referenced in the coordinate frame of the appendage.*

Table 6.
Physical parameters for the satellite, appendages, and PZT used in MATLAB/Simulink simulations[a,b].

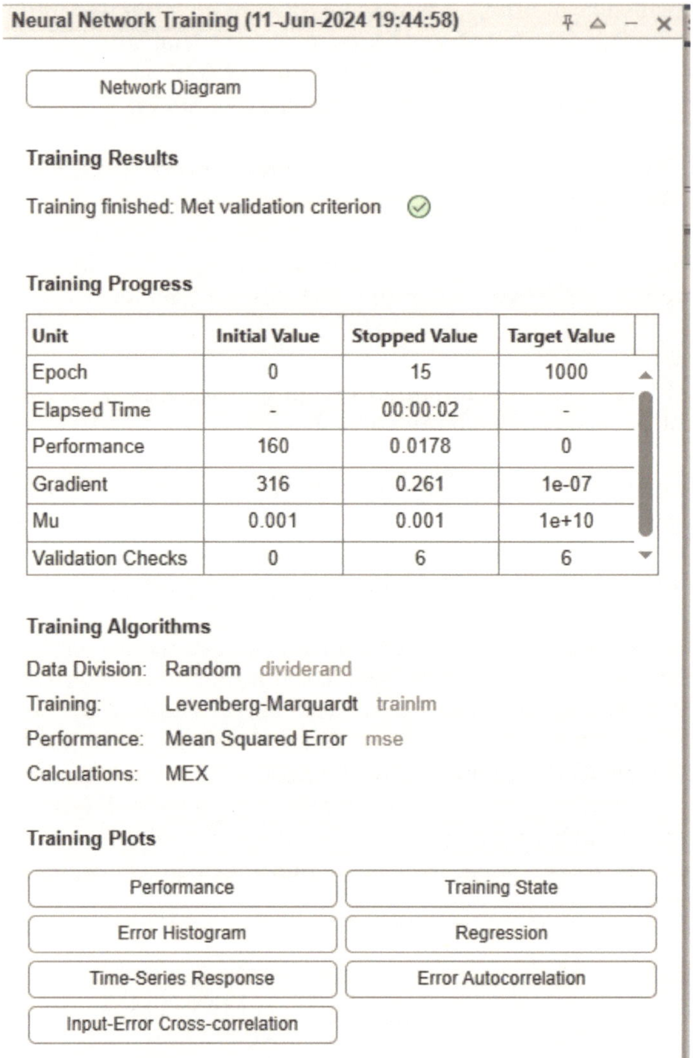

Figure 6.
Progression of neural network training.

Figure 7 displays the time-series response of the first output element. The predicted values closely match the actual data, with minimal deviations observed in the error plot, demonstrating the model's high accuracy.

The error distribution is illustrated in **Figure 8**. The majority of prediction errors are concentrated around zero, indicating that the model performs well without introducing significant bias.

Figure 9 presents a regression plot comparing the predicted outputs against the target values. The high correlation observed confirms a strong agreement between the two, further validating the model's accuracy.

The residual error autocorrelation, shown in **Figure 10**, demonstrates that all significant correlations remain within the confidence limits. This confirms that the model does not exhibit systematic error patterns and has successfully captured the key system dynamics.

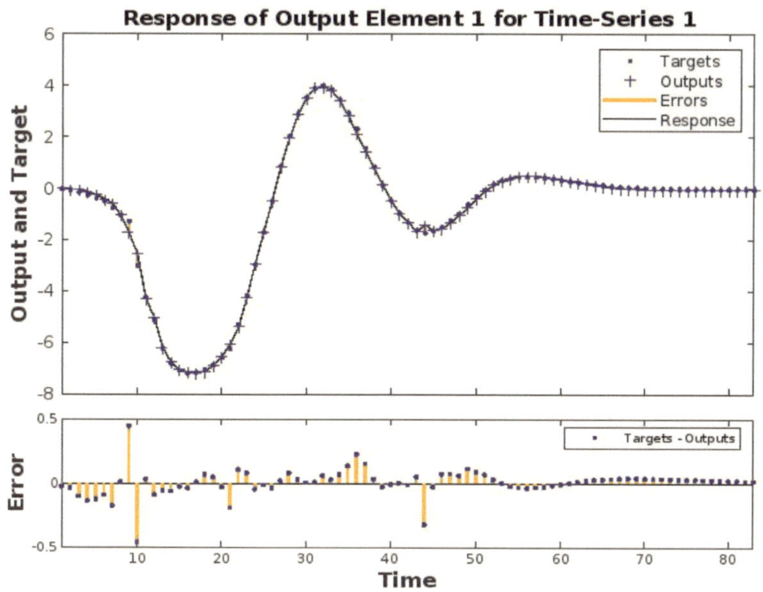

Figure 7.
Time-series output for first element.

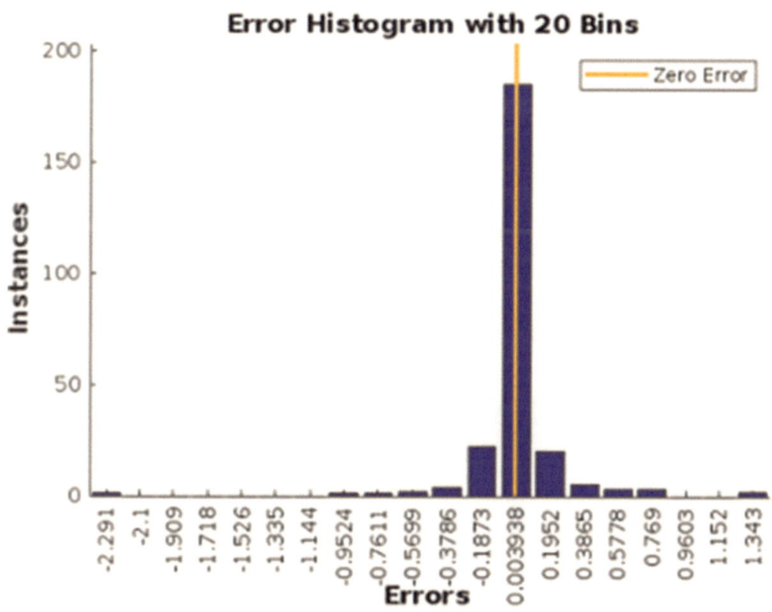

Figure 8.
Distribution of prediction errors.

The cross-correlation between input variables and residual errors is depicted in **Figure 11**. The results indicate that most correlation values remain within the confidence bounds, suggesting that the model's predictions were not significantly influenced by input biases.

The network's training, validation, and test performance, measured in terms of mean squared error (MSE), is summarized in **Figure 12**. The lowest validation error

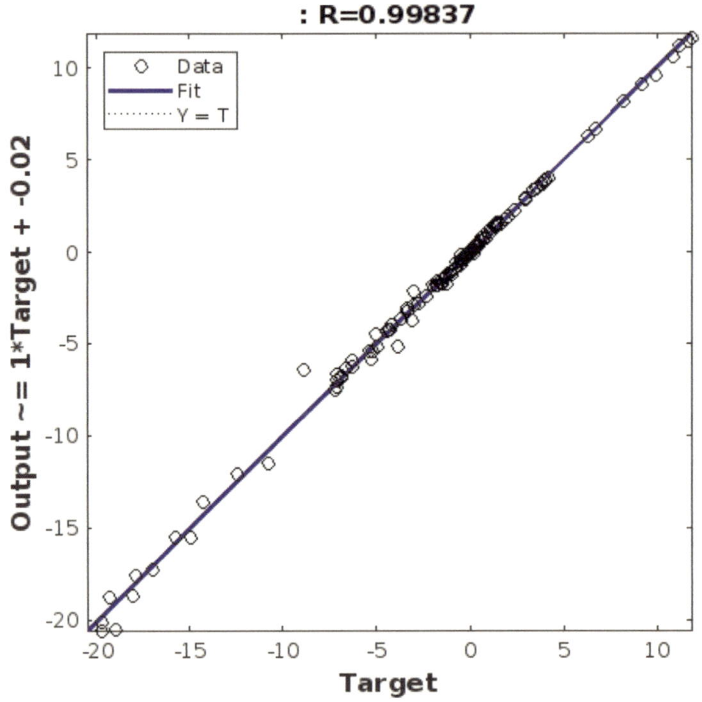

Figure 9.
Regression analysis of predicted vs. target values.

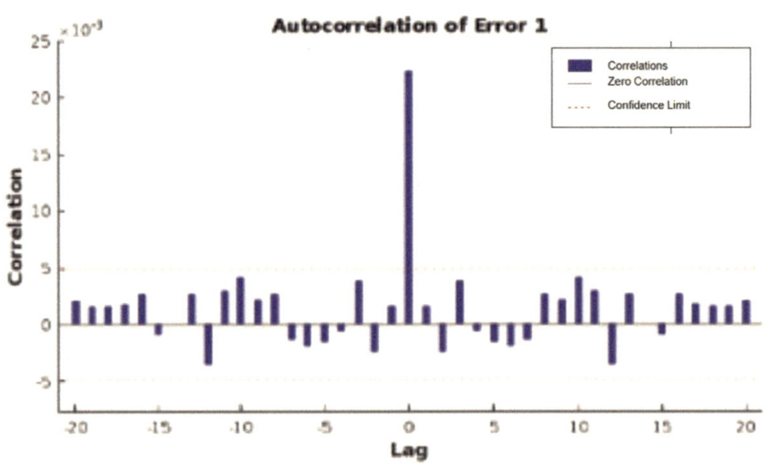

Figure 10.
Autocorrelation of residual errors.

was recorded at epoch 8, with an MSE of 0.03909. The smooth convergence of errors across all datasets confirms the model's ability to generalize effectively to unseen data.

Figure 13 visualizes the evolution of modal time coefficients under model uncertainty and disturbances for both NMPC and NMPC-NARX approaches. Unlike standard NMPC, which relies on predefined dynamic models, the NMPC-NARX framework integrates a neural network-based system identification model that

DOI: http://dx.doi.org/10.5772/intechopen.1010037

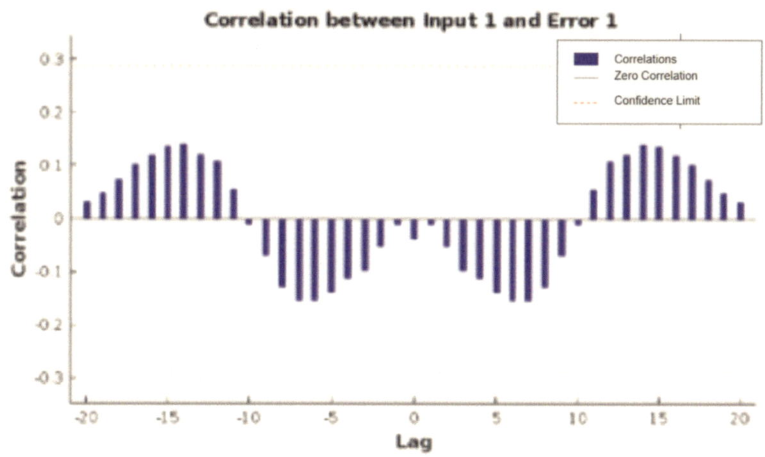

Figure 11.
Cross-correlation between input and residual errors.

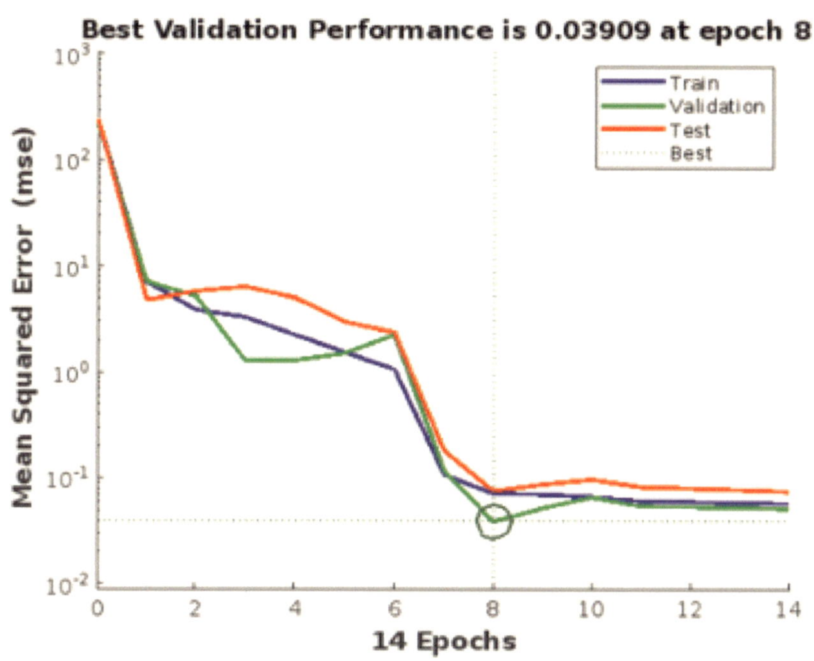

Figure 12.
Training, validation, and test mean squared error (MSE).

correlates applied PZT voltages with resulting vibrations, continuously refining its estimates using a neural network discriminator (NND).

The NMPC-NARX method consistently outperformed conventional NMPC in mitigating vibrations. The modal time coefficient plots reveal a significantly faster decay rate and reduced oscillation amplitudes when NMPC-NARX was employed, demonstrating its superior ability to suppress vibrations effectively. These results highlight the improved resilience and adaptability of the NMPC-NARX approach across various disturbance scenarios.

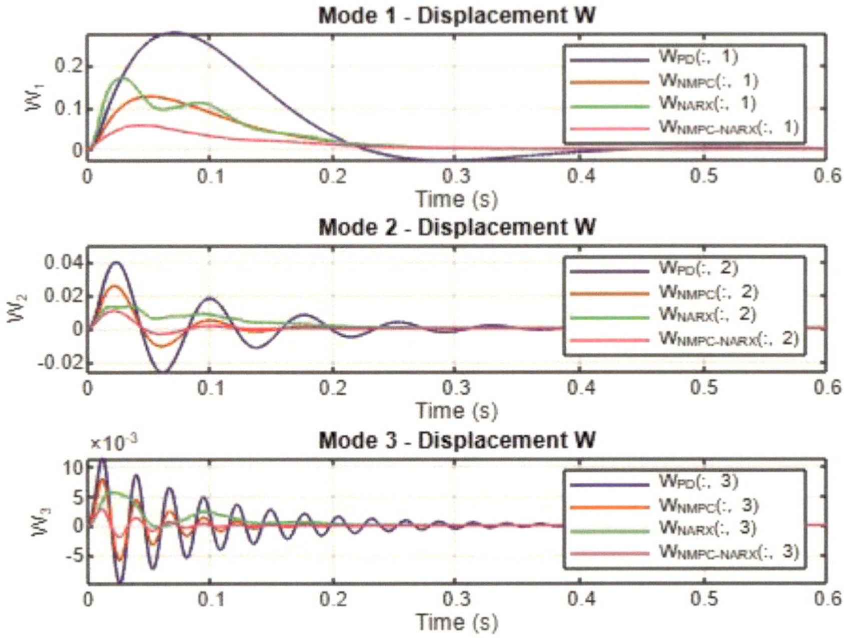

Figure 13.
Comparison of modal time coefficient response for NMPC-NARX, NMPC, and PD control methods.

Figure 14 illustrates the variation of the PZT modal force F_c over time for both NMPC and NMPC-NARX strategies under uncertain conditions. The NMPC-NARX method required lower control efforts while achieving superior vibration suppression

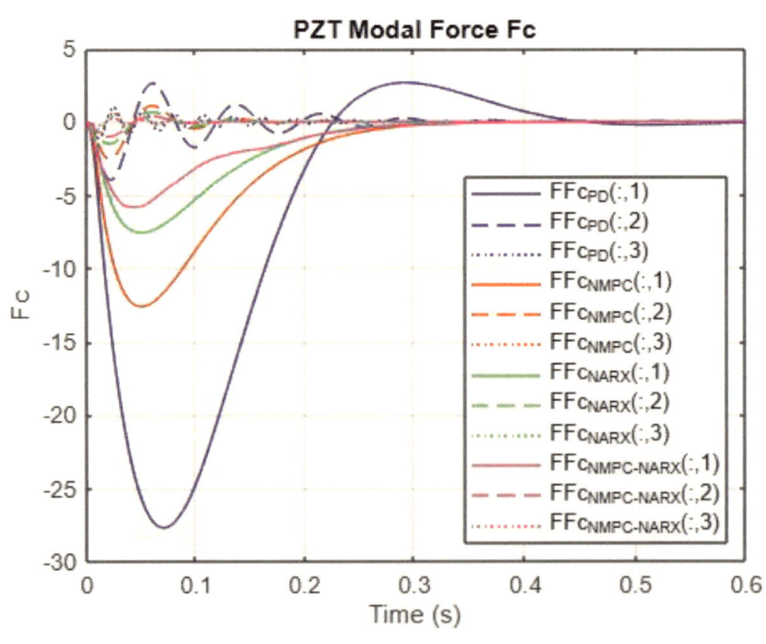

Figure 14.
Time evolution of PZT modal force F_c under NMPC-NARX, NMPC, and PD control strategies.

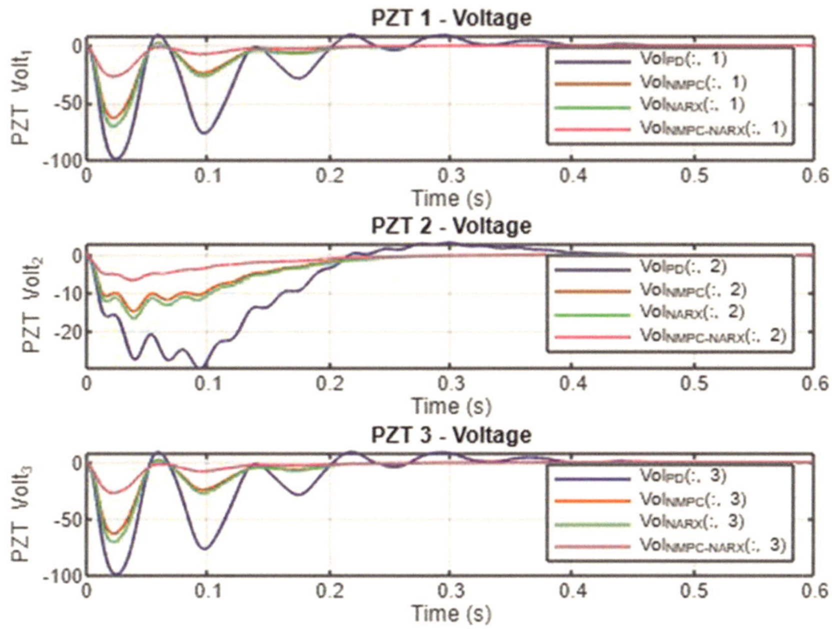

Figure 15.
Control voltage profiles for PZT 1, 2, and 3 under NMPC-NARX, NMPC, and PD control methods.

compared to NMPC. This result reinforces the effectiveness of the NMPC-NARX approach in handling dynamic disturbances with minimized actuation force.

Figure 15 presents the applied control voltages to PZT actuators under NMPC and NMPC-NARX approaches. The NMPC-NARX control strategy consistently utilized lower voltage levels, confirming its capability to maintain effective vibration control with reduced energy expenditure.

4. Conclusion

This chapter introduced *advanced control methodologies* for space systems, integrating *nonlinear model predictive control (NMPC)*, *neural network-based adaptation*, and *passivity-based stability techniques* to address the challenges of coupled *structural-attitude dynamics* in spacecraft equipped with flexible appendages and multiple robotic manipulators. The research presented in this work advances the field of *precision control in space applications*, providing a robust framework for *on-orbit servicing, space debris removal, and autonomous assembly operations*.

A key focus of this study was the development of an *integrated NMPC-NARX control framework* for *vibration suppression in flexible satellite structures*. By formulating a coupled *attitude and structural dynamics model*, the research provided insights into *how spacecraft motion interacts with elastic deformations*. The NMPC approach effectively predicted future system states and optimized control actions, ensuring high-precision regulation of flexible structures. Meanwhile, the *NARX-based state estimation*, enhanced with a *neural network discriminator (NND)*, demonstrated strong adaptability to system parameter variations and external disturbances. Unlike traditional model-based methods, this *learning-based approach* enabled real-time adaptation, making it particularly well-suited for space environments with *high uncertainty and dynamic conditions*.

Beyond vibration suppression, this chapter also explored the *multi-robot control problem* in *space-based robotic systems*. The details of this work are provided in Ref. [37]. The NMPC-based multi-robot coordination framework leveraged predictive optimization to efficiently compute *trajectories and force distribution among manipulators*, ensuring stable and synchronized motion control. The proposed control methodology was demonstrated to enhance *task efficiency* while maintaining *actuator constraints*, making it highly applicable to complex operations such as *on-orbit servicing, autonomous docking, and space structure assembly*.

A significant contribution of this work was the incorporation of *passivity-based stability analysis* to ensure the robustness of *nonlinear NMPC controllers*. The passivity-based framework provided rigorous *stability guarantees*, particularly in scenarios with *time-varying dynamics, actuator saturation, and unpredictable external disturbances*. By embedding passivity constraints within the NMPC formulation, the control system effectively *regulated energy flow within the system*, preventing instability due to *uncontrolled oscillations* or *excessive control efforts*. This approach is particularly advantageous for *space robotics and flexible appendage control*, where stability considerations are critical due to the inherent *interactions between rigid and flexible components*.

Simulation results throughout the chapter validated the effectiveness of the *integrated NMPC-NARX framework*, demonstrating *superior performance over conventional methods* in handling *model uncertainties, disturbances, and strict actuator constraints*. The results highlighted:

- *Enhanced vibration suppression* through NMPC-NARX, reducing oscillations and improving structural stability.

- *Improved multi-robot coordination* with predictive optimization for *synchronized motion control*.

- *Passivity-based stability guarantees*, ensuring *robust and bounded* system responses under nonlinear and uncertain dynamics.

Table 7 presents a detailed performance evaluation of the integrated NMPC-NARX strategy against conventional NMPC when subjected to model discrepancies and external perturbations. The assessment highlights key aspects such as vibration attenuation efficiency, control force stability, and voltage regulation.

The methodologies developed in this study offer *direct applicability to next-generation space missions*, including *on-orbit servicing, space-based manufacturing, modular space station assembly, and active debris removal*. The successful integration of *predictive control, learning-based adaptation, and passivity constraints* represents a *significant advancement* in the field of *autonomous space robotics and flexible spacecraft control*.

While this research provides a *strong theoretical and simulation-based foundation*, future work should focus on *real-time implementation* using *hardware-in-the-loop simulations* and *experimental validation in microgravity environments*. Additionally, exploring *hardware acceleration techniques* (e.g., FPGA or GPU-based implementations) could enhance the real-time feasibility of NMPC-based control for computationally intensive space applications.

Performance metric	Description
Response of modal time coefficients	
NMPC-NARX	Exhibits rapid vibration suppression with lower oscillation amplitudes in the presence of external disturbances.
NMPC	Moderate damping response, more sensitive to variations in the system model.
PZT modal force (F_c)	
NMPC-NARX	Produces a smoother force response with a peak amplitude of ± 0.23 N.
NMPC	Displays higher-frequency oscillations with peak force reaching ± 0.28 N.
Control voltage utilization	
NMPC-NARX	Operates with consistently reduced voltage requirements (± 90V), even in uncertain conditions.
NMPC	Requires higher voltages (± 100V) with noticeable oscillatory variations during disturbance rejection.

Table 7.
Performance evaluation of NMPC-NARX and NMPC under model uncertainty and disturbance.

In conclusion, this chapter demonstrates that *integrating NMPC, NARX-based learning, and passivity constraints* enables a highly *adaptive, robust, and scalable control framework for space systems*. These techniques pave the way for *greater autonomy, stability, and efficiency* in space operations, providing *a critical foundation for the future of intelligent spacecraft and robotic missions*.

Author details

Sean Kalaycioglu[1,2*†] and Anton de Ruiter[1†]

1 Toronto Metropolitan University, Toronto, Canada

2 AIMechatroniX Inc., Richmond Hill, Canada

*Address all correspondence to: skalay@torontomu.ca

† These authors contributed equally.

References

[1] Nanos K, Papadopoulos EG. On the dynamics and control of free-floating space manipulator systems in the presence of angular momentum. Frontiers in Robotics and AI. 2017;**4**:26. DOI: 10.3389/frobt.2017.00026

[2] Papadopoulos E, Aghili F, Ma O, Lampariello R. Robotic manipulation and capture in space: A survey. Frontiers in Robotics and AI. 2021;**8**:686723. DOI: 10.3389/frobt.2021.686723

[3] Yoshida K. Engineering test satellite VII flight experiments for space robot dynamics and control: Theories on laboratory test beds ten years ago, now in orbit. International Journal of Robotics Research. 2003;**22**:321-335. DOI: 10.1177/0278364903022005003

[4] Krebs D. OSAM 2 (Archinaut 1). Gunter's Space Page; 2022. Available from: https://space.skyrocket.de/doc_sdat/osam-2.htm [Accessed: April 05, 2025]

[5] NASA. NASA's Robotic OSAM-1 Mission Completes its Critical Design Review. NASA Feature. Washington, D. C.: National Aeronautics and Space Administration; 2022

[6] DLR. Orbital Hub DLR Vision 2025. DLR Institute of Space Systems Analysis Space Segment; 2022. Available from: https://elib.dlr.de/98589/1/Brosch%C3%BCre_24425-Post-ISS_A4_030715_C.pdf

[7] Chatterjee J. Legal issues relating to unauthorised space debris remediation. In: Proceedings of the 65th International Astronautical Congress; Toronto, Canada. 2014. pp. 1-20

[8] Zhao P, Liu J, Wu C. Survey on research and development of on-orbit active debris removal methods. Science China Technological Sciences. 2020;**63**: 2188-2210. DOI: 10.1007/s11431-020-1661-7

[9] Billot C, Ferraris S, Rembala R, Cacciatore F, Tomassini A, Biesbroek RE. Deorbit: Feasibility study for an active debris removal. In: 3rd European Workshop on Space Debris Modeling and Remediation. Paris, France: Centre National d'Etudes Spatiales; 2014

[10] Zhang F, Huang P. Releasing dynamics and stability control of maneuverable tethered space net. IEEE/ASME Transactions on Mechatronics. 2016;**22**:983-993. DOI: 10.1109/tmech.2016.2628052

[11] Forshaw J, Aglietti G, Salmon T, Retat I, Burgess C, Chabot T, et al. The remove debris ADR mission: Preparing for an international space station launch. In: Proceedings of the 7th European Conference on Space Debris; Darmstadt, Germany. 2017

[12] Biesbroek R, Innocenti L, Wolahan A, Serrano SM. e.Deorbit – ESA's active debris removal mission. In: Proceedings of the 7th European Conference on Space Debris; Darmstadt, Germany. 2017

[13] Shuyou Y, Marcus R, Hong C, Frank A. Inherent robustness properties of quasi-infinite horizon nonlinear model predictive control. Automatica. 2014;**50** (9):2269-2280. DOI: 10.1016/j.automatica.2014.07.014

[14] Aghili F. Optimal trajectories and robot control for detumbling a non-cooperative satellite. Journal of Guidance, Control, and Dynamics. 2020; **43**:981-988. DOI: 10.2514/1.g004758

[15] Seweryn K, Basmadji FL, Rybus T. Space robot performance during tangent

capture of an uncontrolled target satellite. The Journal of the Astronautical Sciences. 2022;**69**:1017-1047. DOI: 10.1007/s40295-022-00330-2

[16] Raisi M, Noohian A, Fallah S. A fault-tolerant and robust controller using model predictive path integral control for free-flying space robots. Frontiers in Robotics and AI. 2022;**9**:1027918. DOI: 10.3389/frobt.2022.1027918

[17] Hewing L, Wabersich KP, Menner M, Zeilinger MN. Learning-based model predictive control: Toward safe learning in control. Annual Review of Control, Robotics, and Autonomous Systems. 2020;**3**:269-296. DOI: 10.1146/annurev-control-090419-075625

[18] Rawlings JB. Tutorial overview of model predictive control. IEEE Control Systems. 2000;**20**(3):38-52. DOI: 10.1109/37.845037

[19] Shi Y, Zhang K. Advanced model predictive control framework for autonomous intelligent mechatronic systems: A tutorial overview and perspectives. Annual Reviews in Control. 2021;**51**:76-89. DOI: 10.1016/j.arcontrol.2021.05.002

[20] Christofides P, Scattolini D, Muñoz de la Peña D, Liu J. Distributed model predictive control: A tutorial review and future research directions. Computers and Chemical Engineering. 2013;**51**:21-41. DOI: 10.1016/j.compchemeng.2012.05.011

[21] Kalaycioglu S, Misra KA. Approximate solutions for vibrations of deploying appendages. Journal of Guidance, Control, and Dynamics. 1991; **14**(2):287-293

[22] Morato MM, Normey-Rico JE, Sename O. Model predictive control design for linear parameter varying systems: A survey. Annual Reviews in Control. 2020;**49**:64-80. DOI: 10.1016/j.arcontrol.2020.04.003

[23] Mayne DQ, Rawlings CV, Rao CV, Scokaert POM. Constrained model predictive control: Stability and optimality. Automatica. 2000;**36**:789-814. DOI: 10.1016/S0005-1098(99)00214-9

[24] Kalaycioglu S, de Ruiter A. Dual arm coordination of redundant space manipulators mounted on a spacecraft. Robotica. 2023;**41**(8):2489-2518. DOI: 10.1017/S0263574723000504

[25] Psomiadis E, Papadopoulos E. Model-based/model predictive control design for free floating space manipulator systems. In: Proceedings of the IEEE 30th Mediterranean Conference on Control and Automation (MED). Vouliagmeni, Greece: Macmillan Inc; 2022

[26] Rybus R, Seweryn K, Sasiadek J. Application of predictive control for manipulator mounted on a satellite. Archives of Control Sciences. 2018;**28**(1):105-118. DOI: 10.24425/122989

[27] Kalaycioglu S, Giray M, Asmer H. Vibration control of flexible manipulator using smart structures. In: Proceedings of the 12th IEEE International Symposium on Intelligent Control. 1997. pp. 415-420

[28] Vukov M, Gros S, Horn G, Frison G, Geebelen K, Jørgensen JB, et al. Real-time nonlinear MPC and MHE for a large-scale mechatronic application. Control Engineering Practice. 2015;**45**:64-78. DOI: 10.1016/j.conengprac.2015.08.012

[29] Englert T, Volz A, Mesmer F, Rhein S, Graichen K. A software framework for embedded nonlinear model predictive

control using a gradient-based augmented Lagrangian approach (GRAMPC). Optimization and Engineering. 2019;**20**(3):769-809. DOI: 10.1007/s11081-018-09408-7

[30] Rathai KMM. Synthesis and real-time implementation of parameterized NMPC schemes for automotive semi-active suspension systems [PhD thesis]. Grenoble, France: Grenoble INP, Communauté Université Grenoble Alpes; 2020

[31] Quirynen R, Vukov M, Zanon M, Diehl M. Autogenerating microsecond solvers for nonlinear MPC: A tutorial using ACADO integrators. Optimal Control Applications and Methods. 2015; **36**(5):685-704. DOI: 10.1002/oca.2151

[32] Wang M, Luo J, Walter U. A non-linear model predictive controller with obstacle avoidance for a space robot. Advances in Space Research. 2016;**57**(8): 1737-1746. DOI: 10.1016/j.asr.2015.12.020

[33] Jadbabaie A, Yu J, Hauser J. Stabilizing receding horizon control of nonlinear systems: A control Lyapunov function approach. In: Proceedings of the American Control Conference; San Diego, CA, USA. 1999. pp. 1535-1539

[34] Wang H, Xie Y. Passivity based attitude control of rigid bodies. Asian Journal of Control. 2014;**16**:802-817. DOI: 10.1002/asjc.616

[35] van der Schaft A. L2-Gain and Passivity Techniques in Nonlinear Control. London: Springer; 2000. DOI: 10.1007/978-1-4471-0796-3

[36] Raff T, Ebenbauer C, Allgower F. Nonlinear model predictive control: A passivity-based approach. In: Assessment and Future Directions of Nonlinear Model Predictive Control, Lecture Notes in Control and Information Sciences. Vol. 358. Berlin, Heidelberg: Springer; 2007. pp. 151-162. DOI: 10.1007/978-3-540-72699-9_11

[37] Kalaycioglu S, de Ruiter A. Passivity based nonlinear model predictive control (PNMPC) of multi-robot systems for space applications. Frontiers in Robotics and AI. 2023;**10**:1181128. DOI: 10.3389/frobt.2023.1181128

[38] Kalaycioglu S, de Ruiter A. Coordinated motion and force control of multi-rover robotics system with Mecanum wheels. In: Proceedings of the 2022 IEEE International IoT, Electronics and Mechatronics Conference (IEMTRONICS). Piscataway, NJ: Institute of Electrical and Electronics Engineers (IEEE); 2022. pp. 1-9. DOI: 10.1109/IEMTRONICS55184.2022.9795804

[39] Necsulescu DS, Kim B, Kalaycioglu S. Free motion, collision avoidance and contact motion control for mobile robots. In: Proceedings of the Seventh International IFAC Symposium on Information Control Problems in Manufacturing Technology. Amsterdam, Netherlands: Elsevier Science Publishers; 1993. pp. 223-228. DOI: 10.1016/B978-0-08-041897-1.50042-0

[40] Necsulescu N, Kim B, Kalaycioglu S. Contact motion control for mobile robots. In: Seventh IFAC Symposium on Information Control Problems. Toronto: John Wiley & Sons, Inc.; 1992

[41] Fierro R, Lewis FL. Control of a nonholonomic mobile robot: Backstepping kinematics into dynamics. In: Proceedings of the 34th IEEE Conference on Decision and Control. New Orleans, LA: IEEE; 1995. pp. 3805-3810. DOI: 10.1109/CDC.1995.479190

[42] Tian Y, Sidek N, Sarkar N. Modeling and control of a nonholonomic wheeled

mobile robot with wheel slip dynamics. In: 2009 IEEE Symposium on Computational Intelligence in Control and Automation (CICA). Vol. 2009. New York, NY: IEEE. pp. 7-14. DOI: 10.1109/CICA.2009.4982776

[43] Amengonu YH, Kakad YP. Dynamics and control for constrained multibody systems modeled with Maggi's equation: Application to differential mobile robots part II. IOP Conference Series: Materials Science and Engineering. 2014;**65**:012018. DOI: 10.1088/1757-899X/65/1/012018

[44] Campion G, d'Andrea-Novel B, Bastin G. Controllability and State Feedback Stabilizability of Nonholonomic Mechanical Systems. Berlin, Germany: Springer-Verlag; 1991

[45] Bloch AM, NH MC. Control of mechanical systems with classical nonholonomic constraints. In: Proceedings of the 28th IEEE Conference on Decision and Control. Tampa, FL: IEEE; 1989. pp. 201-205. DOI: 10.1109/CDC.1989.70103

[46] Kalaycioglu S. Control of multiple robot manipulators with optimal force distribution. In: IEEE Canadian Conference on Electrical and Computer Engineering. 1991

[47] Shi Y, Zhang K. Advanced model predictive control framework for autonomous intelligent mechatronic systems: A tutorial overview and perspectives. Annual Reviews in Control. 2021;**52**:170-196. DOI: 10.1016/j.arcontrol.2021.10.008

[48] Ellis M, Durand H, Christofides PD. A tutorial review of economic model predictive control methods. Journal of Process Control. 2014;**24**(8):1156-1178. DOI: 10.1016/j.jprocont.2014.03.010

[49] Michael F. Implementation of linear model predictive control – Tutorial. arXiv preprint arXiv:2109.11986. 2021

[50] Yu S, Reble M, Chen H, Allgöwer F. Inherent robustness properties of quasi-infinite horizon MPC. IFAC Proceedings Volumes. 2011;**44**(1):179-184. DOI: 10.3182/20110828-6-IT-1002.01969

[51] Wei H, Shen C, Shi Y. Distributed Lyapunov-based model predictive formation tracking control for autonomous underwater vehicles subject to disturbances. IEEE Transactions on Systems, Man, and Cybernetics: Systems. 2021;**51**(8):5198-5208. DOI: 10.1109/TSMC.2019.2946127

[52] Wei H, Sun Q, Chen J, Shi Y. Robust distributed model predictive platooning control for heterogeneous autonomous surface vehicles. Control Engineering Practice. 2021;**107**:104655. DOI: 10.1016/j.conengprac.2020.104655

[53] Zhang K, Sun Q, Shi Y. Trajectory tracking control of autonomous ground vehicles using adaptive learning MPC. IEEE Transactions on Neural Networks and Learning Systems. 2021;**32**(12):5554-5564. DOI: 10.1109/TNNLS.2020.3048305

[54] Zou Y, Su X, Li S, Niu Y, Li D. Event-triggered distributed predictive control for asynchronous coordination of multi-agent systems. Automatica. 2019;**99**:92-98. DOI: 10.1016/j.automatica.2018.10.019

[55] Zhang K, Shi Y. Adaptive model predictive control for a class of constrained linear systems with parametric uncertainties. Automatica. 2020;**117**:108974. DOI: 10.1016/j.automatica.2020.108974

[56] Ladoiye JS, Necsulescu DS, Sasiadek J. Force control of surgical robot with

time delay using model predictive control. In: Proceedings of the 15th International Conference on Informatics in Control, Automation and Robotics. Setúbal, Portugal: SCITEPRESS - Science and Technology Publications; 2018. pp. 202-210. DOI: 10.5220/0006908602020210

[57] Gangapersaud RA, Liu G, de Ruiter AHJ. Detumbling of a non-cooperative target with unknown inertial parameters using a space robot. Advances in Space Research. 2019;**63**(12):3900-3915. DOI: 10.1016/j.asr.2019.03.002

[58] Englert T, Völz A, Mesmer F, Rhein S, Graichen K. A software framework for embedded nonlinear model predictive control using a gradient-based augmented Lagrangian approach (GRAMPC). Optimization and Engineering. 2019;**20**(3):769-809. DOI: 10.1007/s11081-018-9417-2

[59] Schubert M, Perfetto S, Dafnis A, Mayer D, Atzrodt H. Development and design of multifunctional lightweight structures for satellite applications. In: Proceedings of the 69th International Astronautical Congress (IAC); Bremen, Germany. 2018

[60] Agnes G. ACT-10-0043 Precision Deployable Mast for the Swot KARIN Instrument. Jet Propulsion Laboratory (JPL); 2014. Available from: https://esto.nasa.gov/forum/estf2014/abstracts/Agnes.htm [Accessed: April 05, 2025]

[61] Angeletti F, Gasbarri P, Sabatini M, Iannelli P. Design and performance assessment of a distributed vibration suppression system of a large flexible antenna during attitude manoeuvres. Acta Astronautica. 2020;**2020**:1-15. DOI: 10.1016/j.actaastro.2020.04.015

[62] Wang ZH, Xu M, Jia YH, Xu SJ, Tang L. Vibration suppression-based attitude control for flexible spacecraft. Aerospace Science and Technology. 2017;**70**:487-496

[63] Wang J, Li D. Experiments study on attitude coupling control method for flexible spacecraft. Acta Astronautica. 2018;**147**:393-402

[64] da Fonseca IM, Bainum PM, Da Silva AR. Structural control interaction for an LSS attitude control system using thrusters and reaction wheels. Acta Astronautica. 2007;**60**:865-872

[65] Cheng G, Ianculescu J, Ly J, Kim M. Control-structure interaction study for the space station solar dynamic power module. In: Proceedings of the 30th IEEE Conference on Decision and Control; Brighton, UK. 1991. pp. 2219-2223

[66] Allegranza C, Gaillard L, Le Letty R, Patti S, Scolamiero L, Toso M. Actuators for space applications: State of the art and new technologies. In: 14th International Conference on New Actuators; Bremen, Germany. 2014

[67] Hagood NW, Von Flotow A. Damping of structural vibrations with piezoelectric materials and passive electrical networks. Journal of Sound and Vibration. 1991;**146**(2):243-268

[68] Prasad SE, Wallace JB, Giray M, Kalaycioglu S. Development of composite structures for static shape control. Proceedings of the SPIE – The International Society for Optical Engineering. 1996;**3321**:462-470

[69] Wallace JB, Pettit BE, Prasad SE, Kalaycioglu S, Giray M. Shape control using embedded piezoelectric actuators. IEEE International Symposium on Applications of Ferroelectrics. 1996;**2**:919-922

[70] Angeletti F, Gasbarri P, Sabatini M, Iannelli P. End-to-end design of a robust

attitude control and vibration suppression system for large space smart structures. Acta Astronautica. 2021;**187**: 416-428

[71] Zhong Q, He F, Zhang C. Active vibration control in space frame structures using PZT-based actuators. Journal of Intelligent Material Systems and Structures. 2016;**27**:567-580

[72] Kalaycioglu S. Identification using smart structures. In: Proceedings of the 7th AAS/AIAA Space Flight Mechanics Meeting, Huntsville, AL, February 10-12, 1997. American Astronautical Society; 1997

[73] Forshaw JL, Retat I, Aglietti G, et al. The active space debris removal mission remove debris. Part 1: From concept to launch. Acta Astronautica. 2020;**168**: 293-309. DOI: 10.1016/j.actaastro.2019.12.033

[74] Kalaycioglu S, Silva D. Minimization of vibration of spacecraft appendages during shape control using smart structures. Journal of Guidance, Dynamics and Control - AIAA. 2000;**23** (3):558-561

[75] Koszewnik A. The active vibration control of the plate structure by using LQG controller and piezo-stripes. In: Proceedings of the 22nd International Conference on Methods and Models in Automation and Robotics (MMAR), 28–31 Aug, 2017, Miedzyzdroje, Poland. IEEE; 2017

[76] Yuan Q, Liu Y, Qi N. Active vibration suppression for maneuvering spacecraft with high flexible appendages. Acta Astronautica. 2017; **139**:512-520

[77] Du Y, Wang C, Zhou Y, Lu J. Vibration control for flexible spacecraft using multi-impulse robust input shaper and optimal control method. Journal of Vibration and Control. 2020;**12**(3):797-814

[78] Du Y, Wang C. Dynamic coupling and control of flexible space robots. International Journal of Humanoid Robotics. 2020;**17**(6):2050021

[79] Yang C, Wang Q, Lu W, Li Y. Integrated uncertain optimal design strategy for truss configuration and attitude–vibration control in rigid–flexible coupling structure with interval uncertainties. Nonlinear Dynamics. 2024;**113**(3):2215-2238

[80] Yang C. Interval Riccati equation-based and non-probabilistic dynamic reliability-constrained multi-objective optimal vibration control with multi-source uncertainties. Journal of Sound and Vibration. 2025;**595**:Article 118742

[81] Tokhi MO, Hossain MA. A unified adaptive active control mechanism for noise cancellation and vibration suppression. Journal of Mechanical Systems and Signal Processing. 1996;**10**: 667-682

[82] da Fonseca IM, Rade DA, Goes LCS, de Paula Sales T. Attitude and vibration control of a satellite containing flexible solar arrays by using reaction wheels, and piezoelectric transducers as sensors and actuators. Acta Astronautica. 2017; **139**:357-366. DOI: 10.1016/j.actaastro.2017.07.018

[83] Puig L, Barton A, Rando N. A review on large deployable structures for astrophysics. Acta Astronautica. 2010; **67**:12-26

[84] Kalaycioglu S, Misra KA. Minimization of vibration of axially moving beams. Journal of Canadian Society of Mechanical Engineering. 1989; **13**(4):135-140

[85] Kalaycioglu S, Misra KA. Optimal deployment of spacecraft appendages. Acta Astronautica. 1989;**20**:83-88

[86] Kalaycioglu S. Effect of offset of the point of attachment on the dynamics and stability of spinning flexible appendages. Advances in the Astronautical Sciences. 1987;**65**(Part IL):877-894

[87] Kalaycioglu S, Misra KA. Effects of root-offset of a spacecraft appendage on its dynamics. Journal of Astronautical Sciences. 1990;**38**(4):421-437

[88] Gasbarri P, Monti R, De Angelis C, Sabatini M. Effects of uncertainties and flexible dynamic contributions on the control of a spacecraft full-coupled model. Acta Astronautica. 2014;**94**:515-526

[89] Sabatini M, Gasbarri P, Palmerini GB. Synergetic approach in attitude control of very flexible satellites by means of thrusters and PZT devices. Aerospace Science and Technology. 2020;**96**:2-11

[90] Benanni S, Ankersen F, Arcioni M, et al. Robust attitude control design for the BIOMASS satellite (earth explorer core mission candidate). In: Proceedings of the 18th IFAC Symposium on Automatic Control in Aerospace, 28 August – 2 September, 2011; Milan, Italy. Elsevier; 2011

[91] Iannelli P, Angeletti F, Gasbarri P. Model-based FDI applied to a piezoelectric active vibration suppression system for smart flexible spacecraft. Aerotecnica Missili Spazio. 2021;**100**:147-160

[92] Zhang Y, Li M, Zhang J. Vibration control for rapid attitude stabilization of spacecraft. IEEE Access. 2017;**5**:23502-23510

[93] Zhang J, Guo Z, Zhang Y. Dynamic characteristics of vibration isolation platforms considering the joints of the struts. Acta Astronautica. 2016;**127**:216-225

[94] Xie J, Zhao X, Dong H. Learning-based nonlinear model predictive control with accurate uncertainty compensation. Nonlinear Dynamics. 2021;**104**:3827-3842

[95] He F, Wei H-L, Billings SA. Identification and frequency domain analysis of non-stationary and nonlinear systems using time-varying NARMAX models. International Journal of Systems Science. 2015;**46**(11):2087-2100. DOI: 10.1080/00207721.2013.860202

[96] Song H, Shan X, Hou W, Wang C. An efficient vibration suppression technology of piezoelectric cantilever beam based on the NARX neural network. Mechanics of Advanced Materials and Structures. 2023;**30**(9):2009-2020

[97] Fazzi A, Chiuso A. Data-driven prediction and control for NARX systems. arXiv preprint arXiv:2301.02956. 2023. Available from: https://arxiv.org/abs/2301.02956

[98] Li Z, Zhang Y, Wang FY. A zeroing neural network-based approach to parameter-varying platooning control of connected automated vehicles. IEEE Transactions on Intelligent Vehicles. 2023;**8**(2):1234-1245

[99] Najafqolian MA. Control of aerial robots using convex QP LMPC and learning-based explicit-MPC. IEEE Transactions on Industrial Informatics. 2024;**20**(6):8317-8327

[100] Sun GY, He Z, Xu J, Li F, Xu J. Cooperative model predictive levitation control for two-points electromagnetic levitation system of high-speed maglev vehicle. IEEE Transactions on Intelligent

Vehicles. 2023;**99**:1-12. DOI: 10.1109/TIV.2023.3329073

[101] Kelley J, Hagan MT. Comparison of neural network NARX and NARMAX models for multi-step prediction using simulated and experimental data. Expert Systems with Applications. 2024;**237**: 121437. DOI: 10.1016/j.eswa.2023.121437

[102] Yan J, Deller JR Jr. NARMAX model identification using a set-theoretic evolutionary approach. Signal Processing. 2016;**123**:30-41. DOI: 10.1016/j.sigpro.2015.12.001

[103] Rahrooh A, Shepard S. Identification of nonlinear systems using NARMAX model. Nonlinear Analysis. 2009;**71**:e1198-e1202. DOI: 10.1016/j.na.2009.01.150

[104] Kalaycioglu S, De Ruiter A. Vibration control of satellite antennas via NMPC and NARX neural networks. IEEE Transactions on Aerospace and Electronic Systems. 2025;**61**(3):1234-1245

Chapter 5

Super-Twisting Sliding Mode Controller with Fuzzy Logic-Based Moving Sliding Surface for Electronic Throttle Control

Abdul Kareem, Varuna Kumara, Akshatha Naik, Sookshma Adiga and Farana Mohammed Imran

Abstract

This chapter presents the application of a super-twisting sliding mode control scheme with a fuzzy logic-based moving sliding surface for the control of the electronic throttle of gasoline engines. The logic behind this approach is to make use of a moving sliding surface function, in which the sliding surface slope is modified online continuously using a fuzzy logic controller in order to move the sliding surface in the direction of the improved dynamics of the electronic throttle control system. The performance of the proposed approach is evaluated in comparison with the basic super-twisting sliding mode control algorithm with a predetermined sliding surface by conducting MATLAB/SIMULINK-based simulations on the electronic throttle control system model. The results confirm that the proposed control approach exhibits improved dynamic performance in comparison with the basic super-twisting sliding mode controller with a predetermined sliding surface.

Keywords: sliding mode controller, super-twisting control, fuzzy logic, moving sliding surface, electronic throttle

1. Introduction

An electronic throttle control system serves as the link between the pedal of the accelerator and the throttle used for regulating the amount of the entry of air into gasoline engines [1–4]. The basic structure of an electronic throttle system is presented in **Figure 1**. The system's core is a control unit that calculates the target opening angle, which represents the desired valve position of the throttle. This angle is determined based on the torque demand, which depends on the accelerator pedal position, engine speed, and vehicle speed. The goal of the throttle controller is to make sure that the valve's opening angle θ follows the target opening angle r, using a DC motor to achieve this.

The numerous uncertainties and the nonlinearities pose a challenging task to the control of the throttle valve. This makes the super-twisting sliding mode control

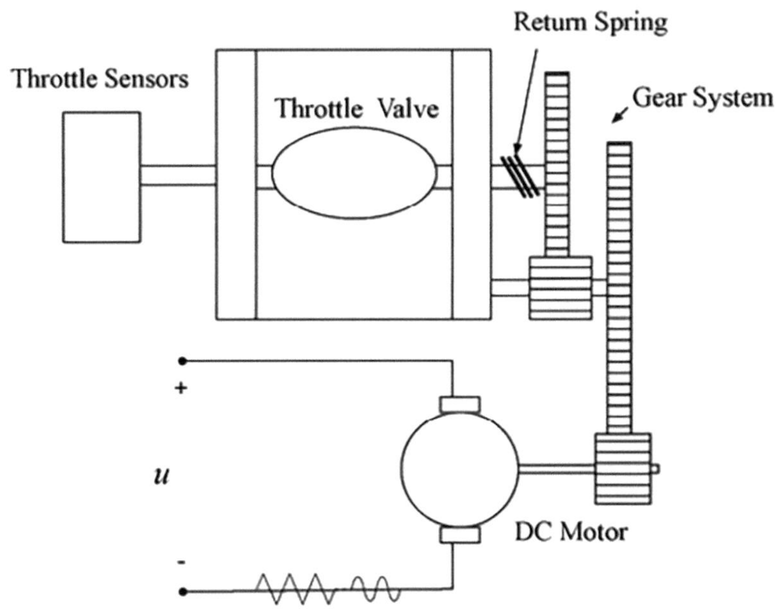

Figure 1.
Electronic throttle system.

scheme a viable option for electronic throttle control [5–20]. This chapter is the extended version of the paper presented at the IEEE International Conference on Fuzzy Systems (FUZZ-IEEE), 2016 [21]. Here, a super-twisting sliding mode control scheme with a fuzzy logic-based moving sliding surface is proposed for electronic throttle control. The controller is evaluated by comparing its performance with the basic super-twisting sliding mode controller with a predetermined sliding surface proposed by Horn and Reichhartinger [4] for the electronic throttle control.

2. Super-twisting sliding mode approach with fuzzy logic control-based moving sliding surface

The basic premise of the moving sliding surface stems from the fact that a super-twisting sliding mode controller with a lower value of sliding surface slope results in slower tracking dynamics and hence, slower error convergence, whereas a super-twisting sliding mode controller with a higher value of sliding surface slope results in faster dynamics and hence, faster error convergence, but at the cost of degrading tracking accuracy. Therefore, there is a compromise between error convergence time and tracking accuracy [22–24]. This can be resolved by the movement of the sliding surface of the super-twisting sliding mode controller as presented in **Figure 2**.

The movement of the sliding surface can be obtained by adjusting the value of the sliding surface slope online based on the values of the error variables, error in the output $e(t)$, and rate of change of error in the output $\dot{e}(t)$, with a requirement that the sliding surface slope must be positive for making sure that the system under control is stable. There is no exact mathematical model of the relationship between the error variables and the sliding surface slope. Hence, the updating of the sliding surface slope

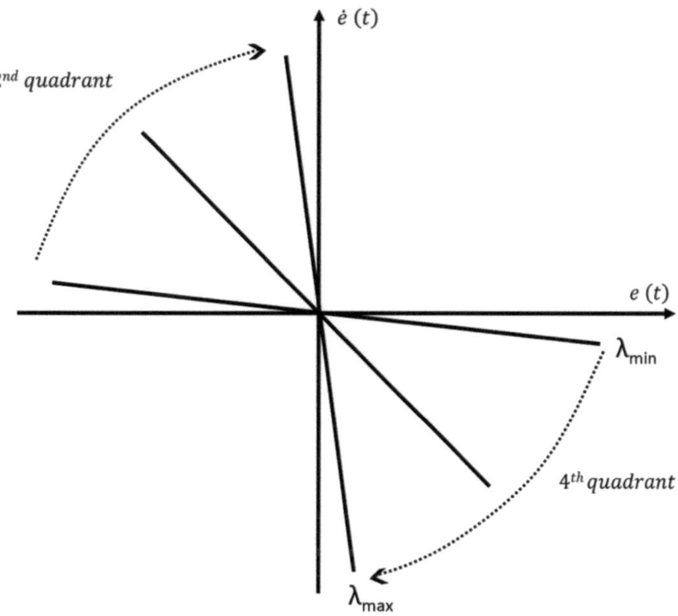

Figure 2.
Movement of sliding surface.

can be achieved by a fuzzy logic inference system with a two-dimensional rule base designed based on the approximate rules derived from expert knowledge [21, 25–31]. The rules take the form "If $e(t)$ is e_f and $\dot{e}(t)$ is e_f, then $\ddot{e}(t)$ is Lambda," where e_f, e_f, and Lambda are the fuzzy sets of $e(t)$, $\dot{e}(t)$, and $\ddot{e}(t)$, respectively.

The error in the output $e(t)$ and rate of change of error in the output $\dot{e}(t)$ are converted to normalized error $E(t)$ and normalized rate of change of error $\dot{E}(t)$, respectively, in the unit range of [−1, 1] before applying them to the fuzzy logic controller (FLC). The output of the FLC, which is the normalized value of the slope in the range [0,1] is scaled by an appropriate output scaling factor, giving the sliding surface slope of the super-twisting sliding mode controller. The inputs of the fuzzy logic controller may take positive or negative values. However, the output of the FLC must always be positive to ensure stability. The inputs are therefore selected within the range of [−1,1], while the output is within [0,1]. The input membership functions include negative big (NB), negative medium (NM), negative small (NS), zero (ZE), positive small (PS), positive medium (PM), and positive big (PB), as depicted in **Figure 3**. For the output, the membership functions are categorized as very very small (VVS), very small (VS), small (S), medium (M), big (B), very big (VB), and very very big (VVB), as illustrated in **Figure 4**.

The rules of the FLC are designed such that the sliding surface is rotated in such a way that the performance of the electronic throttle system is improved. From **Figure 2**, it is clear that the sliding surface should move in the same direction as that of the system when the representative point is in the second or fourth quadrants. This can be achieved by the rule base as follows:

If error $\dot{E}$ (t) is Positive Medium (PM) and the rate of change of error E(t) is Positive Big (PB), then the slope of the sliding surface $\lambda(t)$ is Small (S).

If error $\dot{E}$ (t) is Positive Medium (PM) and the rate of change of error E(t) is Positive Medium (PM), then the slope of the sliding surface $\lambda(t)$ is Medium (M).

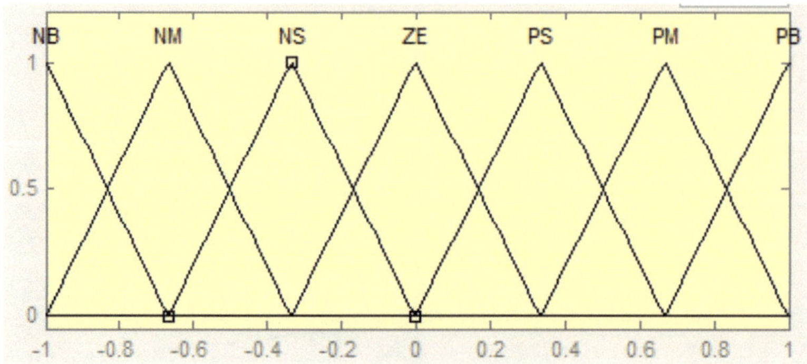

Figure 3.
Input membership functions.

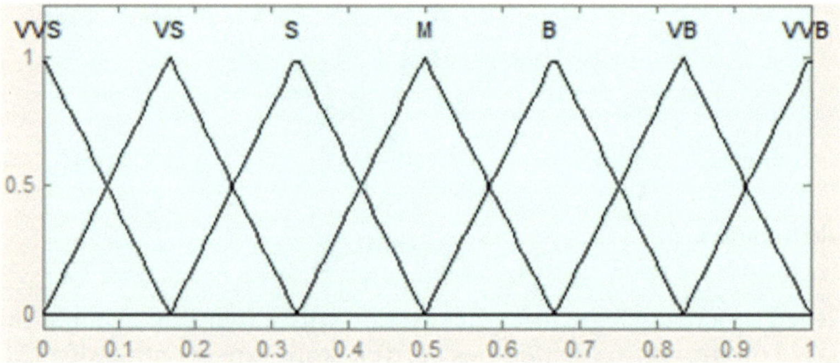

Figure 4.
Output membership functions.

If error $\dot{E}$ (t) is Positive Medium (PM) and the rate of change of error E(t) is Positive Small (PS), then the slope of the sliding surface λ(t) is Big (B).

If error $\dot{E}$ (t) is Positive Medium (PM) and the rate of change of error E(t) is Zero (ZE), then the slope of the sliding surface λ(t) is Very Big (VB).

If error $\dot{E}$ (t) is Positive Medium (PM) and the rate of change of error E(t) is Negative Small (NS), then the slope of the sliding surface λ(t) is Big (B).

If error $\dot{E}$ (t) is Positive Medium (PM) and the rate of change of error E(t) is Negative Medium (NM), then the slope of the sliding surface λ(t) is Medium(M).

If error $\dot{E}$ (t) is Positive Medium (PM) and the rate of change of error E(t) is Negative Big (NB), then the slope of the sliding surface λ(t) is Small (S).

If error $\dot{E}$ (t) is Positive Small (PS) and the rate of change of error E(t) is Positive Big (PB), then the slope of the sliding surface λ(t) is Very Small (VS).

If error $\dot{E}$ (t) is Positive Small (PS) and the rate of change of error E(t) is Positive Medium (PM), then the slope of the sliding surface λ(t) is Small (S).

If error $\dot{E}$ (t) is Positive Small (PS) and the rate of change of error E(t) is Positive Small (PS), then the slope of the sliding surface λ(t) is Medium (S).

If error $\dot{E}$ (t) is Positive Small (PS) and the rate of change of error E(t) is Zero (ZE), then the slope of the sliding surface λ(t) is Big (B).

If error $\dot{E}$ (t) is Positive Small (PS) and the rate of change of error E(t) is Negative Small (NS), then the slope of the sliding surface λ(t) is Medium (M).

If error $\dot{E}$ (t) is Positive Small (PS) and the rate of change of error E(t) is Negative Medium (NM), then the slope of the sliding surface λ(t) is Small (S).

If error $\dot{E}$ (t) is positive Small (PS) and the rate of change of error E(t) is Negative Big (NB), then the slope of the sliding surface λ(t) is Very Small (VS).

If error $\dot{E}$ (t) is Zero (ZE) and the rate of change of error E(t) is Positive Big (PB), then slope of the sliding surface λ(t) is Very Very Small (VVS).

If error $\dot{E}$ (t) is Zero (ZE) and the rate of change of error E(t) is Positive Medium (PM), then the slope of the sliding surface λ(t) is Very Small (VS).

If error $\dot{E}$ (t) is Zero (ZE) and the rate of change of error E(t) is Positive Small (PS), then the slope of the sliding surface λ(t) is Small (S).

If error $\dot{E}$ (t) is Zero (ZE) and the rate of change of error E(t) is Zero (ZE), then the slope of the sliding surface λ(t) is Medium (M).

If error $\dot{E}$ (t) is Zero (ZE) and the rate of change of error E(t) is Negative Small (NS), then the slope of the sliding surface λ(t) is Small (S).

If error $\dot{E}$ (t) is Zero (ZE) and the rate of change of error E(t) is Negative Medium (NM), then the slope of the sliding surface λ(t) is Very Small (VS).

If error $\dot{E}$ (t) is Zero (ZE) and the rate of change of error E(t) is Negative Big (NB), then the slope of the sliding surface λ(t) is Very Very Small (VVS).

If error $\dot{E}$ (t) is Negative Small (NS) and the rate of change of error E(t) is Positive Big (PB), then the slope of the sliding surface λ(t) is Very Small (VS).

If error $\dot{E}$ (t) is Negative Small (NS) and the rate of change of error E(t) is Positive Medium (PM), then the slope of the sliding surface λ(t) is Small (S).

If error $\dot{E}$ (t) is Negative Small (NS) and the rate of change of error E(t) is Positive Small (PS), then the slope of the sliding surface λ(t) is Medium (M).

If error $\dot{E}$ (t) is Negative Small (NS) and the rate of change of error E(t) is Positive Small (PS), then the slope of the sliding surface λ(t) is Big (M).

If error $\dot{E}$ (t) is Negative Small (NS) and the rate of change of error E(t) is Negative Small (NS), then the slope of the sliding surface λ(t) is Medium (M).

If error $\dot{E}$ (t) is Negative Small (NS) and the rate of change of error E(t) is Negative Medium (NM), then the slope of the sliding surface λ(t) is Small (S).

If error $\dot{E}$ (t) is Negative Small (NS) and the rate of change of error E(t) is Negative Big (NB), then slope of the sliding surface λ(t) is Very Small (VS).

If error $\dot{E}$ (t) is Negative Medium (NM) and the rate of change of error E(t) is Positive Big (PB), then the slope of the sliding surface λ(t) is Small (S).

If error $\dot{E}$ (t) is Negative Medium (NM) and the rate of change of error E(t) is Medium (M), then the slope of the sliding surface λ(t) is Medium (M).

If error $\dot{E}$ (t) is Negative Medium (NM) and the rate of change of error E(t) is Positive Small (PS), then the slope of the sliding surface λ(t) is Big (B).

If error $\dot{E}$ (t) is Negative Medium (NM) and the rate of change of error E(t) is Zero (ZE), then the slope of the sliding surface λ(t) is Very Big (VB).

If error $\dot{E}$ (t) is Negative Medium (NM) and the rate of change of error E(t) is Negative Small (NS), then the slope of the sliding surface λ(t) is Big (B).

If error $\dot{E}$ (t) is Negative Medium (NM) and the rate of change of error E(t) is Negative Medium (NM), then the slope of the sliding surface λ(t) is Medium (M).

If error $\dot{E}$ (t) is Negative Medium (NM) and the rate of change of error E(t) is Negative Big (NB), then the slope of the sliding surface λ(t) is Small (S).

If error $\dot{E}$ (t) is Negative Big (NB) and the rate of change of error E(t) is Positive Big (PB), then the slope of the sliding surface λ(t) is Medium (M).

If error $\dot{E}$ (t) is Negative Big (NB) and the rate of change of error E(t) is Positive Big (PB), then the slope of the sliding surface λ(t) is Medium (M).

If error $\dot{E}$ (t) is Negative Big (NB) and the rate of change of error E(t) is Positive Small (PS), then the slope of the sliding surface λ(t) is Very Big (VB).

If error $\dot{E}$ (t) is Negative Big (NB) and the rate of change of error E(t) is Zero (ZE), then the slope of the sliding surface λ(t) is Very Very Big (VVB).

If error $\dot{E}$ (t) is Negative Big (NB) and the rate of change of error E(t) is Negative Small (NS), then the slope of the sliding surface λ(t) is Very Big (VB).

If error $\dot{E}$ (t) is Negative Big (NB) and the rate of change of error E(t) is Negative Medium (NM), then the slope of the sliding surface λ(t) is Big(B).

If error $\dot{E}$ (t) is Positive Big (PB) and the rate of change of error E(t) is Positive Big (PB), then the slope of the sliding surface λ(t) is Medium(M).

If error $\dot{E}$ (t) is Positive Big (PB) and the rate of change of error E(t) is Positive Medium (PM), then the slope of the sliding surface λ(t) is Big(B).

If error $\dot{E}$ (t) is Positive Big (PB) and the rate of change of error E(t) is Positive Small (PS), then the slope of the sliding surface λ(t) is Very Big (VB).

If error $\dot{E}$ (t) is Positive Big (PB) and the rate of change of error E(t) is Zero (Z), then the slope of the sliding surface λ(t) is Very Very Big (VVB).

If error $\dot{E}$ (t) is Positive Big (PB) and the rate of change of error E(t) is Negative Small (NS), then slope of the sliding surface λ(t) is Very Big (VB).

If error $\dot{E}$ (t) is Positive Big (PB) and the rate of change of error E(t) is Negative Medium (NM), then the slope of the sliding surface λ(t) is Big (B).

If error $\dot{E}$ (t) is Positive Big (PB) and the rate of change of error E(t) is Negative Big (NB), then the slope of the sliding surface λ(t) is Medium (M).

The centroid method is suitable for defuzzification as this method is very accurate and gives smooth output [29]. In this method, the crisp output is obtained by computing the centroid of the aggregated fuzzy output as given in Eq. (1). The input-output characteristics of the designed fuzzy logic controller are given in **Figure 5**.

$$z^* = \frac{\int \mu_C(z).z\,dz}{\int \mu_C(z)\,dz} \tag{1}$$

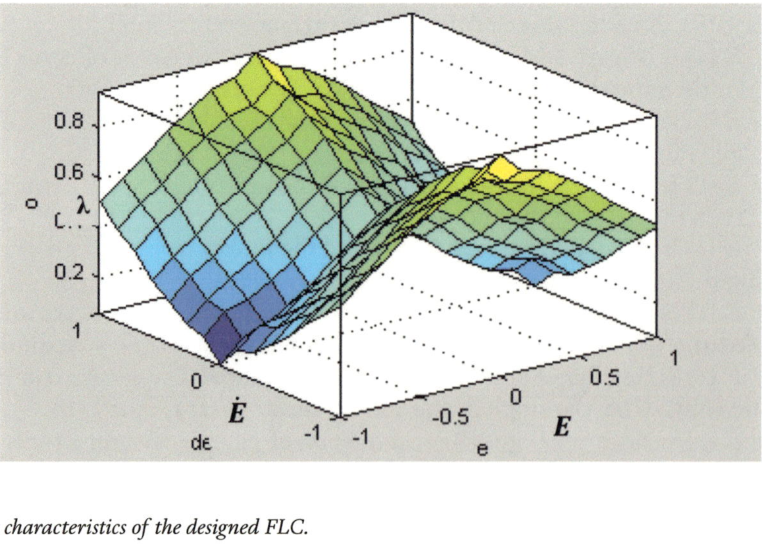

Figure 5.
Input-output characteristics of the designed FLC.

Where, z^* is the defuzzified value, z is the output variable, and $\int \mu_C(z)$ is the membership function of the aggregated fuzzy output.

3. Performance analysis of the proposed controller

The proposed controller is evaluated in comparison with the basic super-twisting sliding mode controller with a predetermined sliding surface proposed by Horn and Reichhartinger [4] for different types of reference signals of the throttle control system given by Eq. (2) [4]. The study is based on MATLAB/SIMULINK-based simulations.

$$\frac{d^2\theta}{dt^2} = -f_t(\theta) - \beta_t \frac{d\theta}{dt} - \delta_t sign\left(\frac{d\theta}{dt}\right) + \gamma_t u \tag{2}$$

Where,

$$f_t(\theta) = \alpha_t(\theta - \theta_0) + \mu_t sign(\theta - \theta_0) \tag{3}$$

where, θ is the opening angle of the valveand u is the input voltage of the dc motor. The nominal values of the parameters of the system model are α_t = 80 s^{-2}, $\beta_t = 100\, s^{-1}$, $\gamma_t = 120$ rad $s^{-2}V^{-1}$, $\delta_t = 70\,V$, μ_t = 150 V and θ_0 =*0.095* rad.

3.1 Case 1: A staircase type reference

The performance comparisons of the system for the proposed controller and the basic super-twisting sliding mode controller with a predetermined sliding surface for a staircase type reference are given in **Figures 6–10**. The integral of absolute error (IAE) and integral of time-absolute error (ITAE) are used as performance indices for the comparison.

The opening angle of the valve for the system with the designed controller and the basic sliding mode controller with a predetermined surface for a stair case type reference are given in **Figure 6**. It is clear that the designed controller exhibits a faster response as compared to the basic sliding mode controller with a predetermined sliding surface. The faster error convergence is evident in the error plot shown in **Figure 7**. The IAE curve and ITAE curve shown in **Figures 8** and **9**, respectively, show that the proposed controller is faster throughout the response. The IAE indices for the designed controller and the basic sliding mode controller with a fixed sliding surface are 0.1346 and 0.3237, respectively, while the ITAE indices are 0.9185 and 2.298, respectively. The enhancement in the dynamic performance of the designed controller is attributed to the adjustment of the sliding surface, as illustrated in **Figure 10**.

3.2 Case 2: A ramp-type reference

The performance comparisons of the system for the designed controller and the basic sliding mode controller with a predetermined sliding surface for a ramp-type reference of **Figure 11** are given in **Figures 12–15**.

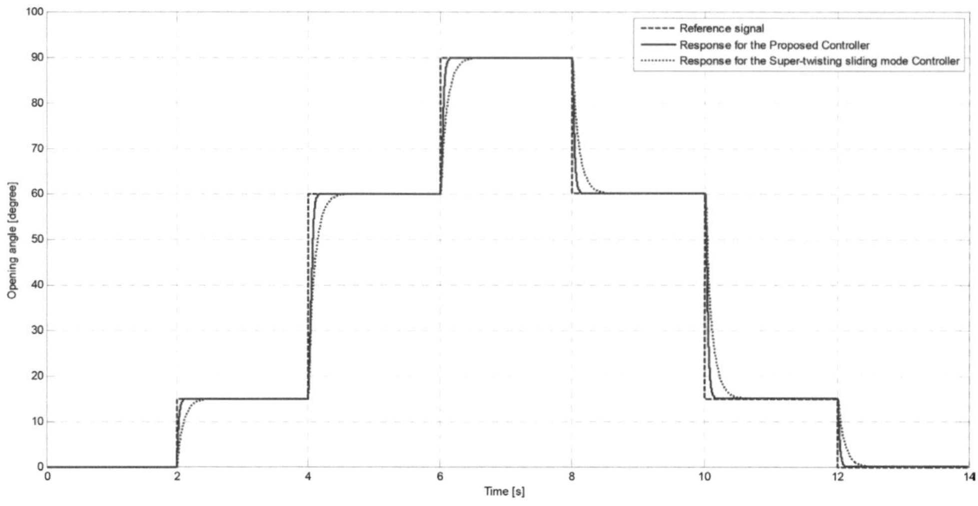

Figure 6.
Throttle valve opening angle for a staircase type reference.

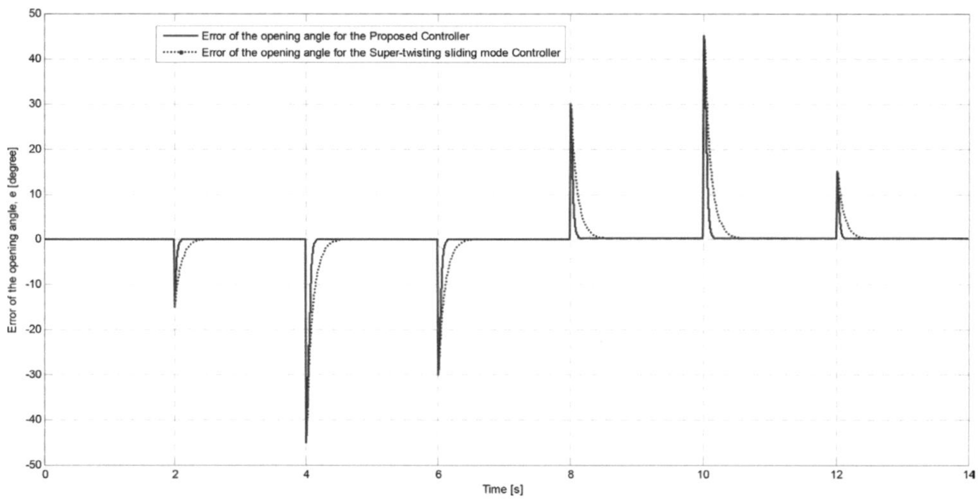

Figure 7.
The error of the opening angle for a staircase type reference.

The output response plots of the opening angle appear quite similar for the ramp-type reference. Therefore, error curves are utilized for a more precise analysis. As illustrated in **Figure 12**, the error curves demonstrate that the designed controller produces a smaller error compared to the basic super-twisting sliding mode controller with a predetermined surface for the ramp-type reference. The maximum errors for the designed controller and the basic super-twisting sliding mode controller with a fixed sliding surface are 0.054° and 0.47°, respectively. The IAE and ITAE curves in **Figures 13** and **14**, respectively, indicate that the designed controller consistently results in lower error throughout the response compared to the basic super-twisting sliding mode controller with a fixed surface. The IAE indices for the designed

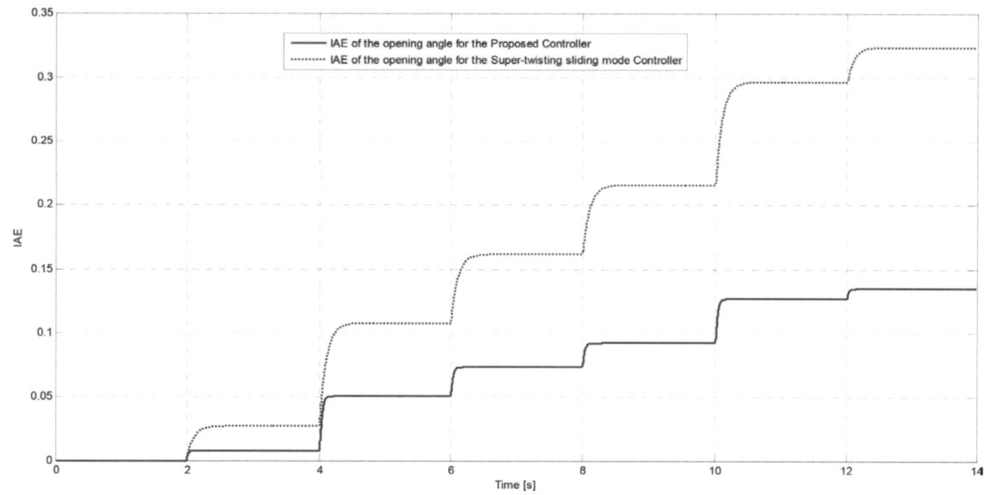

Figure 8.
IAE of the throttle valve opening angle for a staircase type reference.

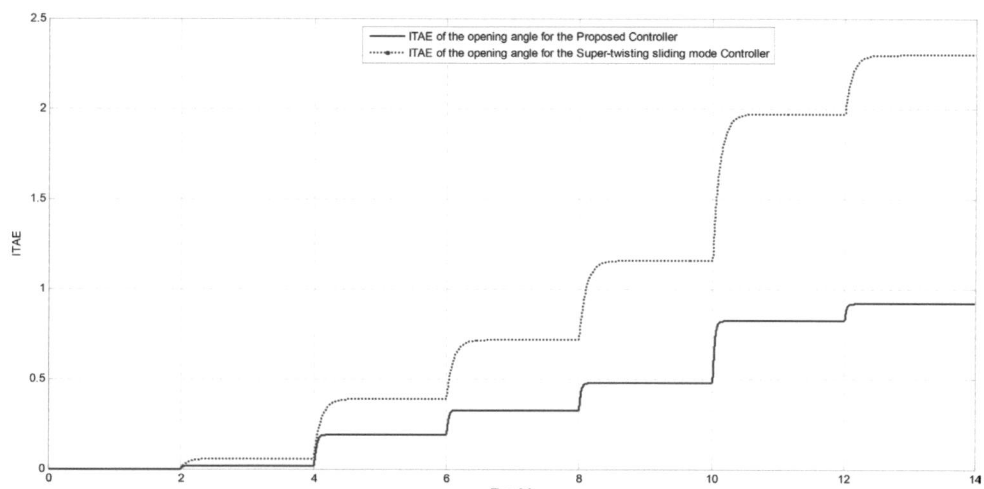

Figure 9.
ITAE of the opening angle for a staircase type reference.

controller and the basic super-twisting sliding mode controller with a predetermined surface are 0.00015 and 0.0035, respectively, while the ITAE indices are 0.00087 and 0.020, respectively. The enhanced dynamic performance of the designed controller can be attributed to the adjustment of the sliding surface, as depicted in **Figure 15**.

3.3 Case 3: A complex type reference

The performance comparisons of the system for the proposed controller and the basic super-twisting sliding mode controller with a predetermined sliding surface for a complex case type reference are given in **Figures 16–20**.

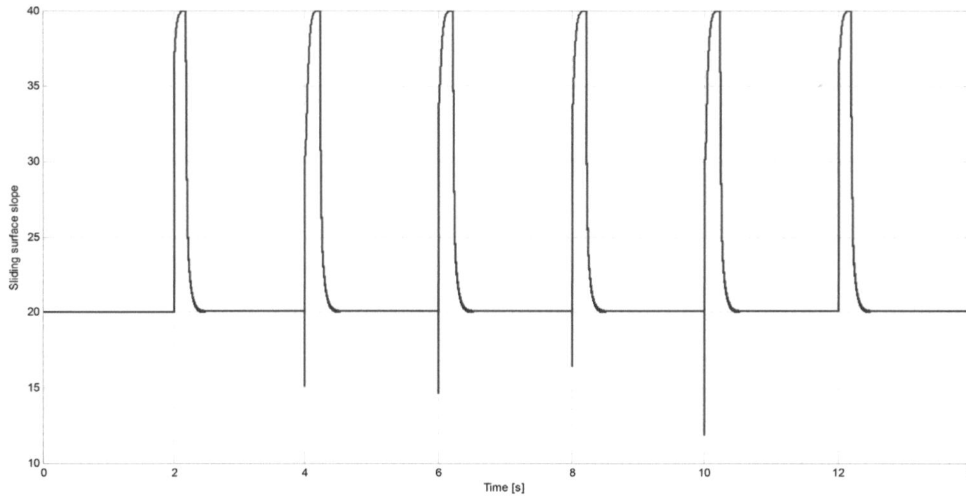

Figure 10.
Sliding surface slope of the designed controller for a staircase type reference.

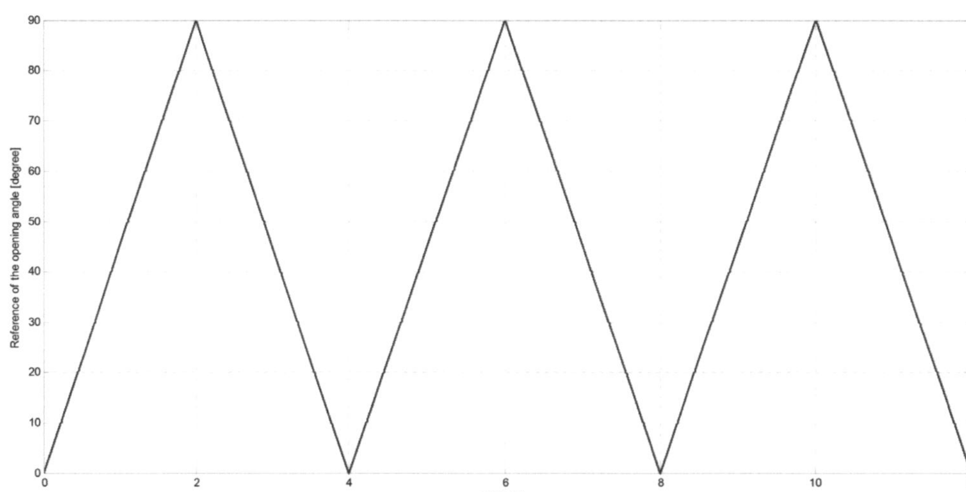

Figure 11.
A ramp-type reference.

The output response plots for the opening angle appear quite similar for this reference. Therefore, error curves are utilized for a more detailed analysis. As shown in **Figure 17**, the proposed controller produces a smaller error compared to the basic super-twisting sliding mode controller with a predetermined sliding surface. The maximum errors are respectively 0.052° and 0.47° for the designed controller and the basic super-twisting sliding mode controller with a predetermined surface. The IAE and ITAE curves shown in **Figures 18** and **19**, respectively show that the error is lesser throughout the response for the designed controller as compared to the basic super-twisting sliding mode controller with a predetermined sliding surface. The IAE

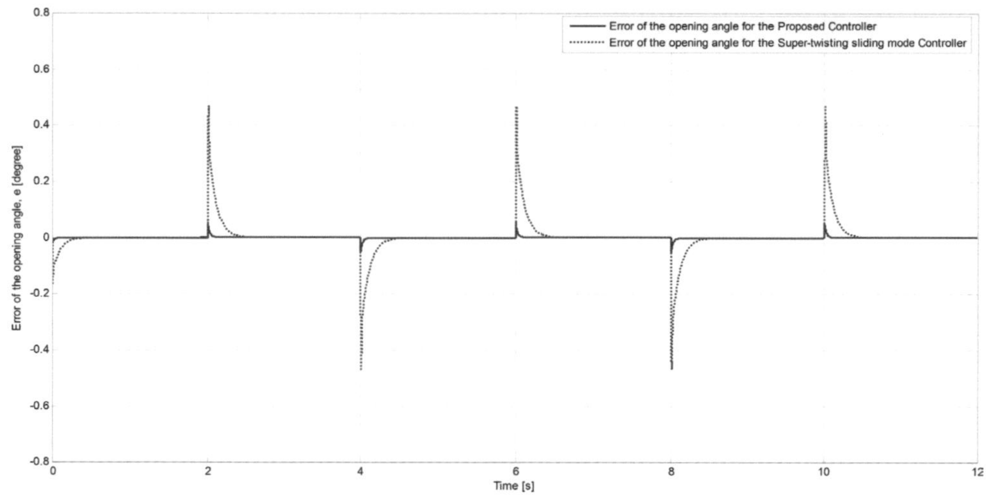

Figure 12.
The error of the throttle valve opening angle for a ramp-type reference.

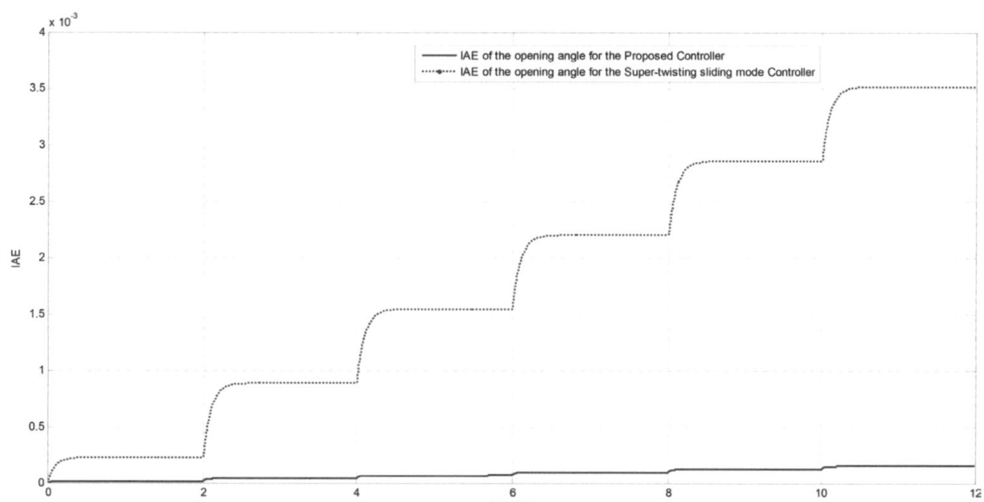

Figure 13.
IAE of the throttle valve opening angle for a ramp-type reference.

indices for the designed controller and the basic super-twisting sliding mode controller with a fixed sliding surface are $8.9 \times 10^{\wedge}(-5)$ and 0.0018, respectively, while the ITAE indices are 0.0004 and 0.0081, respectively. The enhancement in the dynamic performance of the designed controller is attributed to the adjustment of the sliding surface, as illustrated in **Figure 20**.

From the simulation studies, it is clear that the designed controller exhibits better performance in terms of improved dynamic response as compared to the basic super-twisting sliding mode controller with a predetermined sliding surface.

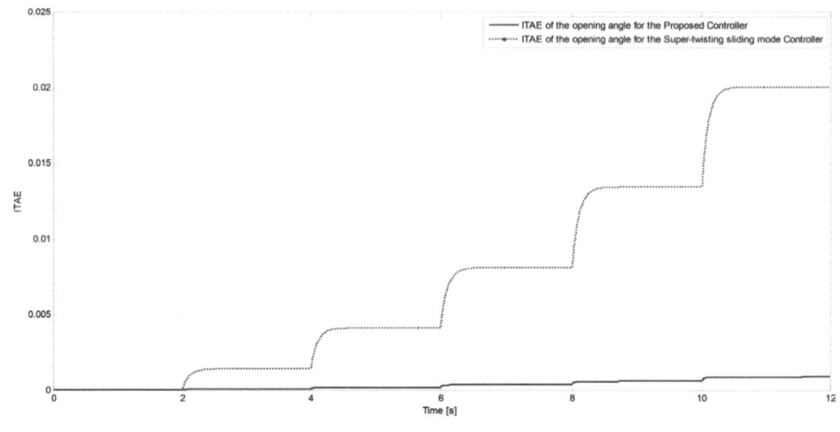

Figure 14.
ITAE of the opening angle for a ramp-type reference.

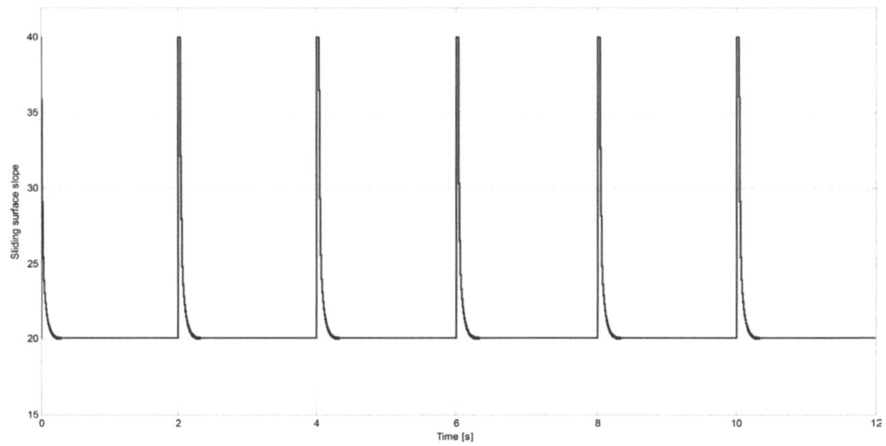

Figure 15.
Sliding surface slope of the designed controller for a ramp-type reference.

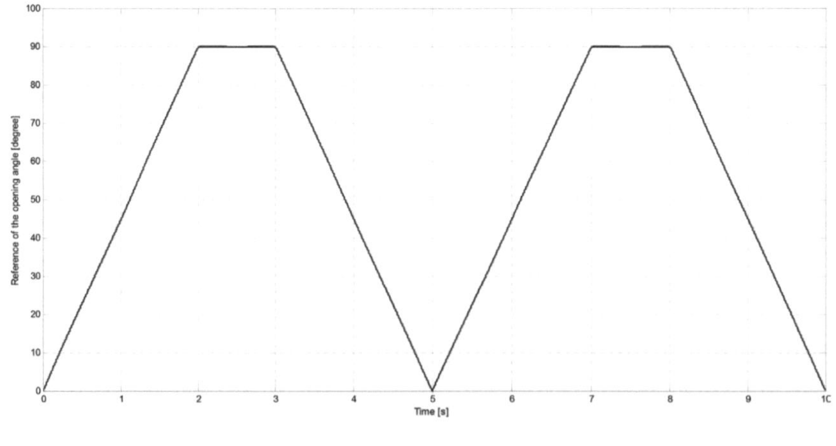

Figure 16.
A complex type reference.

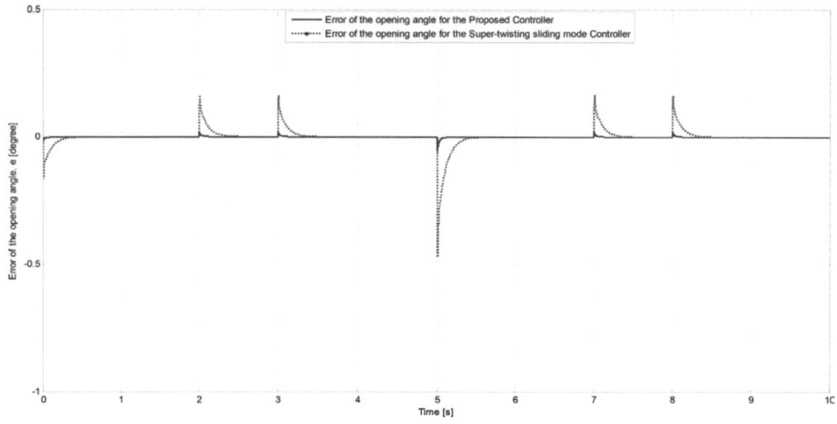

Figure 17.
The error of the opening angle for a complex type reference.

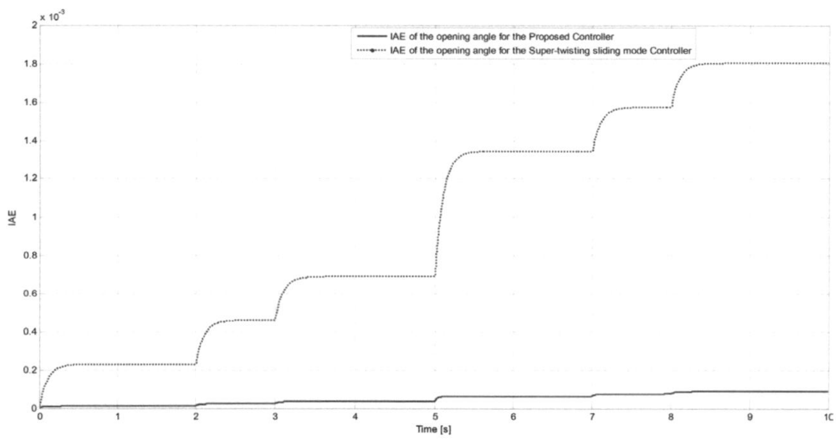

Figure 18.
IAE of the opening angle for a complex type reference.

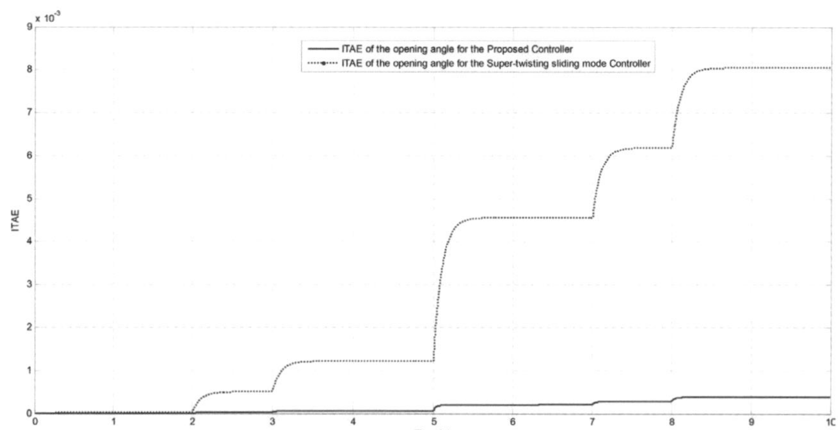

Figure 19.
ITAE of the opening angle for a complex reference.

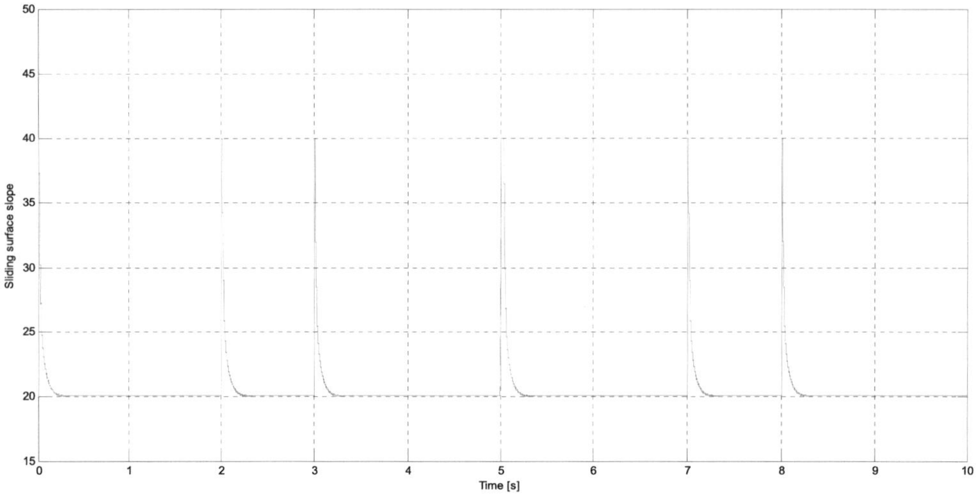

Figure 20.
Sliding surface slope of the designed controller for a complex type reference.

4. Conclusion

This chapter introduces the application of a super-twisting sliding mode control strategy combined with a fuzzy logic-based moving sliding surface for electronic throttle control. The concept behind this approach involves utilizing a time-varying sliding surface function, where the slope of the surface is continuously adjusted online by a fuzzy logic controller. This adjustment is designed to steer the sliding surface toward enhancing the dynamic performance of the electronic throttle control system. The performance of the proposed control approach is evaluated in comparison with the basic super-twisting sliding mode control algorithm with a predetermined sliding surface by conducting MATLAB/SIMULINK simulations on the electronic throttle control system model. The results confirm that the proposed control approach exhibits an improved dynamic performance in comparison with the basic super-twisting sliding mode controller with a predetermined sliding surface.

Author details

Abdul Kareem*, Varuna Kumara, Akshatha Naik, Sookshma Adiga and Farana Mohammed Imran
Moodlakatte Institute of Technology, Kundapura, Karnataka, India

*Address all correspondence to: afthabakareem@gmail.com

References

[1] Meetam K, Gonsrang S, Dechwayukul C, Kaewapichai W, Neranon P, Vittayaphadung N, et al. Modeling and tuning of electronic throttle control system in formula student car. International Journal of Automotive and Mechanical Engineering. 2023;**20**:10917-10930. DOI: 10.15282/ijame.20.4.2023.09.0844

[2] Youssef O, Shalaby RO. Ptimizing automotive electronic throttle control with a modified grey wolf algorithm. In: 5th Novel Intelligent and Leading Emerging Sciences Conference (NILES); 21-23 October. Giza, Egypt: IEEE; 2023. pp. 222-227

[3] Gao J, Feng K, Wang Y, Wu Y, Chen H. Design, implementation and experimental verification of a compensator-based triple-step model reference controller for an automotive electronic throttle. Control Engineering Practice. 2020;**100**:104447. DOI: 10.1016/j.conengprac.2020.104447

[4] Reichhartinger M, Horn M. Application of higher order sliding-mode concepts to a throttle actuator for gasoline engines. IEEE Transactions on Industrial Electronics. 2009;**56**(9):3322-3329. DOI: 10.1109/TIE.2009.2026382

[5] Alt B, Svaricek F. Robust control design for automotive applications: A variable structure control approach. In: Challenges and Paradigms in Applied Robust Control. Rijeka, Croatia: InTech; 2011. pp. 73-92. DOI: 10.5772/16724

[6] Bartolini G, Ferrara A, Usai E, Utkin V. On multi-input chattering-free second-order sliding mode control. IEEE Transactions on Automatic Control. 2000;**45**(9):1711-1717. DOI: 10.1109/9.880629

[7] Bartolini G, Pisano A, Usai E. An improved second-order sliding mode control sheme robust against the measurement noise. IEEE Transactions on Automatic Control. 2004;**49**(10):1731-1736. DOI: 10.1109/TAC.2004.835395

[8] Bartolini G, Pisano A, Usai E. On the second-order sliding mode control of nonlinear systems with uncertain control direction. Automatica. 2009;**45**(12):2982-2985. DOI: 10.1016/j.automatica.2009.09.018

[9] Bartolini G, Elisabetta P. Second order sliding mode control of systems with nonlinear friction. In: Proceedings of the 39th IEEE Conference on Decision and Control; December 12-15 2000. Sydney, Australia: IEEE; 2002. pp. 2192-2197

[10] Boiko I, Fridman L, Pisano A, Usai E. Analysis of chattering in systems with second-order sliding modes. IEEE Transactions on Automatic Control. 2007;**52**(11):2085-2102. DOI: 10.1109/TAC.2007.908319

[11] Boiko I, Fridman L, Pissano A, Usai E. A comprehensive analysis of chattering in second order sliding mode control systems. Lecture Notes in Control and Information Sciences. 2008;**375**:23-49. DOI: 10.1007/978-3-540-79016-7_2

[12] Capisani LM, Ferrara A, Magnani L. Design and experimental validation of a second-order sliding-mode motion controller for robot manipulators. International Journal of Control. 2009;**82**:365-377. DOI: 10.1080/00207170802112591

[13] Chiang HK, Lin WB, Chang YC, Fang CC. Super-twisting second-order sliding mode control for a synchronous reluctance motor. Artificial Life and

Robotics. 2011;**16**:307-310. DOI: 10.1007/s10015-011-0935-7

[14] Choi SB, Park DW, Jayasuriya S. A time-varying sliding surface for fast and robust tracking control of second-order uncertain systems. Automatica. 1994;**30**:899-904. DOI: 10.1016/0005-1098(94)90180-5

[15] Davila A, Moreno JA, Fridman L. Optimal Lyapunov function selection for reaching time estimation of super-twisting algorithm. In: Joint 48th IEEE Conference on Decision and Control and 28th Chines Control Conference (CDC); 15-18 December 2009. Shangai, China: IEEE; 2010. pp. 8405-8410

[16] Derafa L, Fridman L, Benallegue A, Ouldali A. Super-twisting control algorithm for four rotors helicopter attitude tracking problem. In: 11th International Workshop on Variable Structure Systems (VSS); 26-28 June 2010. Mexico City, Mexico: IEEE; 2010. pp. 62-67

[17] Fridman LM. An averaging approach to chattering. IEEE Transactions on Automatic Control. 2001;**46**:1260-1265. DOI: 10.1109/9.940930

[18] Gonzalez T, Moreno JA, Fridman L. Variable gain super-twisting sliding mode control. IEEE Transactions on Automatic Control. 2012;**57**:2100-2105. DOI: 10.1109/TAC.2011.2179878

[19] Durmaz B, Özgören MK, Salamci MU. Sliding mode control for non-linear systems with adaptive sliding surfaces. Transactions of the Institute of Measurement and Control. 2012;**34**(1):56-90. DOI: 10.1177/0142331210384648

[20] Efe MO, Kaynak O, Wilamowski BM. Stable training of computationally intelligent systems by using variable structure systems technique. IEEE Transactions on Industrial Electronics. 2000;**47**:486-496. DOI: 10.1109/41.836365

[21] Azeem MF, Kareem A. A fuzzy logic based super-twisting sliding mode control scheme for electronic throttle control. In: Proceedings of 2016 IEEE International Conference on Fuzzy Systems (FUZZ-IEEE); 24-29 July 2016. Vancouver, BC, Canada: IEEE; 2016. pp. 485-492. DOI: 10.1109/FUZZ-IEEE.2016.7737726

[22] Eksin I, Guzelkaya M, Tokat S. Sliding surface slope adjustment in fuzzy sliding mode controllers. In: Proceedings of the 10th Mediterranean Conference on Control and Automation(MED 2002); 9-12 July 2002. Lisbon, Portugal: IEEE; 2002. pp. 160-168

[23] Eksin I, Guzelkaya M, Tokat S. Self-tuning mechanism for sliding surface slope adjustment. Proceedings of the Institution of Mechanical Engineers Part I: Journal of Systems and Control Engineering (ITU). 2002:393-406

[24] Fallaha CJ, Saad M, Kanaan HY, Al-Haddad K. Sliding-mode robot control with exponential reaching law. IEEE Transactions on Industrial Electronics. 2011;**58**:600-610. DOI: 10.1109/TIE.2010.2045995

[25] Feng G. A survey on analysis and design of model-based fuzzy control systems. IEEE Transactions on Fuzzy Systems. 2006;**4**:676-697. DOI: 10.1109/TFUZZ.2006.883415

[26] Feng G. Analysis and Synthesis of Fuzzy Control Systems: A Model-Based Approach. 1st ed. United States: CRC press; 2010. 299 P. DOI: 10.1201/EBK1420092646

[27] Kareem A, Azeem MF. A novel soft computing based algorithm for

the control of dynamic uncertain systems–An application to DC-DC converters. International Journal of Artificial Intelligence & Applications. 2011;**2**(2):21-30

[28] Fnaiech MA, Betin F, Capolino GA, Fnaiech F. Fuzzy logic and sliding-mode controls applied to six-phase induction machine with open phases. IEEE Transactions on Industrial Electronics. 2010;**57**:354-364. DOI: 10.1109/TIE.2009.2034285

[29] Hellendoorn H, Thomas C. Defuzzification in fuzzy controllers. Journal of Intelligent and Fuzzy Systems. 1993;**1**:109-123. DOI: 10.3233/IFS-1993-1202

[30] Kareem A, Azeem MF. A novel fuzzy logic based adaptive super-twisting sliding mode control algorithm for dynamic uncertain systems. International Journal of Artificial Intelligence & Applications. 2012;**3**(4):21-34. DOI: 10.5121/ijaia.2012.3402

[31] Kareem A. Super-twisting sliding mode controller with fuzzy logic based moving sliding surface for electronic throttle control. International Journal of Advanced Mechatronic Systems. 2017;**7**(3):174-182. DOI: 10.1504/IJAMECHS.2017.086211

Chapter 6

Performance Evaluation of PI and ST-SMC Controllers in Low Voltage/Power DC-Microgrids

Mohamed Amine Hartani, Aissa Benhammou and Abdallah Laidi

Abstract

This work compares four control strategies for a photovoltaic-diesel generator-hybrid energy storage system, focusing on classical proportional-integral (PI) versus nonlinear super-twisting-sliding-mode-control (ST-SMC). The performance is evaluated using root mean square error (RMSE) and mean absolute error (MAE) metrics. The PI controller provides a better DC-Bus voltage response under pulse conditions, achieving an RMSE of 2.48 and MAE of 0.61. In contrast, the ST-SMC excels in battery current tracking, with an RMSE of 27.13 and MAE of 7.10, leading to reductions of 85% in RMSE and 81% in MAE compared to PI. For supercapacitor current control, ST-SMC shows an RMSE of 6.79 and MAE of 0.96, significantly outperforming PI's RMSE of 139.62 and MAE of 24.88, resulting in reductions of 95% and 96%, respectively. Both controllers perform similarly under load mismatch conditions, with ST-SMC slightly better at MAE (0.07 vs. 0.08). Overall system power losses yield similar results for both controllers (RMSE ~14.58, MAE ~9.33). Such results reflect the strength of ST-SMC, regarding robustness and precision, hence quite suitable for high-precision challenges. Considering the novelty of outstanding improvements in current control of the battery and SCs, the ST-SMC is a robust option that fits widely for high stability and reliability systems.

Keywords: classical control, nonlinear control, energy management strategies, stationary DC-microgrids, performance assessment

1. Introduction

Recent statistics show that isolated networks integrate more renewable energy (RE) with energy storage systems (ESS). This is done to reduce electricity concerns in the developed and developing nations [1].

- *United States:* The US DOE reported that, as of 2023, more than 2 GW of standalone ESS capacity has been deployed, with significant contributions from islanded zones like Hawaii and Puerto Rico [2]. Hawaii, in particular, aims to achieve 100% RE by 2045 and has already integrated large PV and ESS into its grid [3].

- *Europe:* The island regions, such as the Canary and Greek islands, have been leading in integrating RE and ESS. By 2024, these regions aim to increase their RE capacity by 50%, supported by EU funding and policies aimed at reducing dependence on fossil fuels. In the Greek islands, the combination of wind (WT), PV, and ESS has resulted in a 30% reduction in diesel fuel consumption for power generation in 2023 [4].

- *Australia:* Islanded zones, such as Tasmania and King Island, have made substantial progress. Tasmania has achieved nearly 100% RE penetration with hydroelectric power complemented by WT/PV projects. King Island, which integrates WT, PV, and ESS, reported a RE penetration of over 65% in 2023 [5].

- *Africa:* In sub-Saharan Africa, countries like Kenya, Tanzania, and Ghana are increasingly deploying standalone RE and ESS in remote and islanded zones. By 2023, over 100 MW of mini-grid systems incorporating PV and ESS have been installed across these regions, significantly improving energy access and reliability. Madagascar's Nosy Be Island has seen a substantial upgrade with a PV-plus-ESS, which has reduced reliance on diesel generators by 70%.

- *Asia:* Indonesia, with its vast archipelago, is focusing on PV and ESS to electrify its many islands. By 2024, Indonesia plans to install 500 MW of PV in its islanded zones, with a significant portion already operational and providing reliable electricity to remote communities [6]. The Philippines with islands like Palawan and Mindanao deploy hybrid systems combining PV, WT, and ESS to reduce power outages and lower energy costs. The country aims for a 50% RE share by 2040 [7].

- *South America:* In Brazil, the government is promoting the use of PV and ESSs in the Amazon region [8]. As of 2023, over 200 remote communities have been electrified using standalone RE systems, significantly enhancing the quality of life and economic opportunities. Chile is with the Atacama Desert region integrating PV and ESS to support mining operations and local communities [9].

1.1 Overview of voltage and power levels

As mentioned in **Table 1**, in modern power distribution, various grid configurations, including *macro-grids, mini-grids, micro-grids, nano-grids, and further configurations of power systems*, are implemented to meet different energy demands and geographical conditions.

These systems utilize a range of voltage levels for both DC and AC configurations. Power system efficiency, reliability, and application applicability depend on DC and AC coupling topologies [29].

Macrogrids are enormous power networks that supply electricity to large regions using high-voltage AC (HVAC) transmission lines. Mini- and microgrids, which service communities and industrial complexes, often use PV and wind turbines. Nanonetworks power specific buildings or small loads, even more locally than grids.

Reference	Application examples	Key topology/focus	Year
	Standalone microgrids		
[10] Standalone	Rural electrification	Battery-supercapacitor HESS	2018
[11] Isolated	Renewable incorporation	Dynamic OPF	2023
[12] DC MG	Fuel cell-based MG	Adaptive fuzzy management	2022
	Community microgrids		
[13] Community	Microgrid energy markets	Brooklyn microgrid model	2018
[14, 15] Multi-system	Energy sharing	Multi-energy microgrids	2022
	Resiliency improvement	Mobile energy storage	2023
	Hybrid microgrids		
[16, 17] Hybrid	Diesel-PV-Battery hybrid	Optimize fuel, cost, and emissions	2023
	PV-WTG-BSS-DG systems	Optimal energy management	2021
[18] ON-Grid	PV under shading	Optimal control	2021
	Islanded microgrids		
[19, 20] Islanded	Stochastic OPF correlated load and PV uncertainties		2022
	Wind, load uncertainties		2023
	Facility-based microgrids		
[21] Facility	Rule-based EMS PV and energy storage		2021
[22] PV-Diesel	Secondary frequency regulation dual-driven predictive control		2023
	Clustered/distributed microgrids		
[23–25] Multi-MG	Decentralized scheduling	Distributed robust scheduling	2020
	HESS in DC-microgrids	Advanced filtration and decoupling	2023
	Multi-level EMS	Fuzzy logic control	2023
[26] DC MG	Energy management	Optimal control strategy	2022
	Planning and optimization in microgrids		
[27] Multi-Carrier	Planning and operation	Multi-energy hubs	2020
[28] Distributed	Energy management	Harris Hawk Optimization	2021

Table 1.
Overview of voltage and power levels in implemented systems.

These grids can be installed using DC or AC, each with pros and cons. DC systems are best for combining renewable energy sources and energy storage systems (ESSs), while AC systems are used because they work with existing infrastructure. Photovoltaic panels and energy storage systems (ESS) are directly connected to the DC bus using DC coupling. The DC bus powers DC loads or reverses to AC loads. Benefits of DC coupling include [30]:

- Higher efficiency due to fewer conversion steps.
- Better compatibility with RESs and ESSs.
- Reduced complexity with multiple Distributed Energy Resources (DERs).

AC coupling involves inverting the DC output from RESs to AC and connecting them to the AC bus. Grid-tied systems employ this design since it works with AC infrastructure. The advantages of AC coupling include [30]:

- Easy integration with existing AC distribution networks.
- Flexibility in combining multiple AC sources and loads.
- Easier synchronization with the grid.

1.2 Case studies in renewable energy applications

In various power system applications, control methods play a critical role in enhancing stability, efficiency, and performance. The PI controller is utilized for load-sharing control of PV/diesel generator systems [31] and HESS in remote DC-MG [32].

The Fractional Order Proportional-Integral-Derivative (FOPID) controller is applied to enhance Vdc bus stability for PV and battery systems [33], load-frequency control [34], and damping low-frequency oscillation [35]. FOPID with artificial intelligence (AI) techniques is used for PMSM drive adaptation [36] and chopper-fed DC motor drive control [37]. AI-enabled PID controllers aid in metaheuristics-based tuning [38–41].

The general type-2 FLC-PID controller addresses complex control issues [42], while PID-FLC is employed in speed control of brushless DC motors speed [43]. Comparative studies between PI and STSMC highlight their respective advantages [44], while Advanced Fractional Order Proportional-Integral Controllers with Adaptive Neuro-Fuzzy Inference Systems (FOPI-ANFIS) are used in PWM control for multi-rotor wind turbines [45].

Sliding Mode Control (SMC) is implemented for input current control in DC-DC buck converters [46], HESS in islanded DC microgrids [47], and controlling DFIG under unbalanced grid conditions [48]. SMC with AI enhances DFIG systems [49] and optimizes wind energy control gains [50].

STSMC is beneficial for active/reactive power control in grid-interfaced PV systems [51], direct power control for PV systems [52], and control of UUV trajectory tracking [53]. Adaptive STSMC techniques are applied to PEM fuel cell systems [54] and hypersonic vehicles [55]. STSMC-FOPI is designed for wind energy systems [56].

Linear Quadratic Regulator (LQR) control with Kalman filter observer is used for DC-DC converters [57], adaptive gain LQR-based EMS for HESS [58], and energy savings in DC-DC converters [59]. The ANFIS is employed in hybrid electric vehicles [60, 61], EV kinetic energy exploitation [62, 63], and smart grids [64], .

Genetic Algorithm Optimized ANFIS (GA-ANFIS) controls hybrid microgrid systems [65]. Fuzzy Logic Controllers (FLC) manage HESS in multi-DC-MGs [25], FL-EMS for DC-MGs [66], and EMS for MGs with batteries and hydrogen ESS [67, 68].

AI-enhanced FLC is used for PV systems [69], load-frequency control in two-area systems [70], marine diesel engines [71], and Maximum Power Point Tracking (MPPT) in grid-connected PV systems [72]. FLC-SMC improves DC-link voltage response in wind-driven DFIG systems [73]. Also, Model Predictive Control (MPC) optimizes AC-MGs [74], while MPC-FO-STSMC regulates DC-MG voltage [75].

These varied control methods significantly improve the efficiency and reliability of power systems incorporating RE and ESS technologies.

1.3 Classical control methods

PI control combines proportional and integral actions to correct the error between a desired setpoint and the process variable. The proportional component reacts to the current error, while the integral component accumulates the error over time, aiming to eliminate steady-state errors [32]. The control law for a PI is given by:

$$u(t) = K_p e(t) + K_i \int_0^t e(\tau) d\tau \tag{1}$$

$u(t)$: Control signal, K_p: Proportional gain, K_i: Integral gain, $e(t) = r(t) - y(t)$: Error signal, where $r(t)$ is the reference input and $y(t)$ is the output. Besides, the K_p increases the response speed and reduces the rise time, while K_i: eliminates the steady-state error by integrating the error over time.

1.3.1 Recent advancements and case studies

PI controllers are widely used due to their simplicity and effectiveness in control applications. However, they do have certain constraints/limitations that lead to failures in some scenarios, so possible solutions to address are shown in **Figure 1**.

1.4 Fractional-order proportional-integral (FOPI) controllers

FOPI control extends the classical PI control by including fractional calculus, allowing for more flexible and precise control dynamics. The FO integrator provides an additional degree of freedom in tuning [33, 76, 77]. The control law for an FOPI controller is:

$$u(t) = K_p e(t) + K_i D_t^{-\lambda} e(t) \tag{2}$$

With $D_t^{-\lambda}$ is the fractional integral operator of order λ, and λ is the fractional order (typically $0 < \lambda \leq 1$). Besides, the K_p adjusts the response speed, K_i adjusts the accumulation of error, while λ provides additional tuning flexibility, enhancing the controller's performance in various dynamic conditions.

1.5 Super twisting sliding mode control (ST-SMC)

STSMC is a robust nonlinear control that addresses the chattering problem in traditional SMC. It ensures finite-time convergence and improved disturbance rejection by using a second-order SMC approach [32, 78]. The control law for STSMC is:

$$u(t) = u_{eq}(t) + u_{st}(t) \tag{3}$$

Where $u_{eq}(t)$ is the equivalent control that ensures the system stays on the sliding surface and $u_{st}(t)$ is the super twisting control to bring the system state to the sliding surface. For a system: $\dot{x}(t) = f(x(t)) + g(x(t))\, u(t)$, the STSMC law is:

$$\begin{cases} u_{st}(t) = -\lambda |s(t)|^{1/2} \text{sign}(s(t)) + v(t) \\ \dot{v}(t) = -k\text{sign}(s(t)) \end{cases} \tag{4}$$

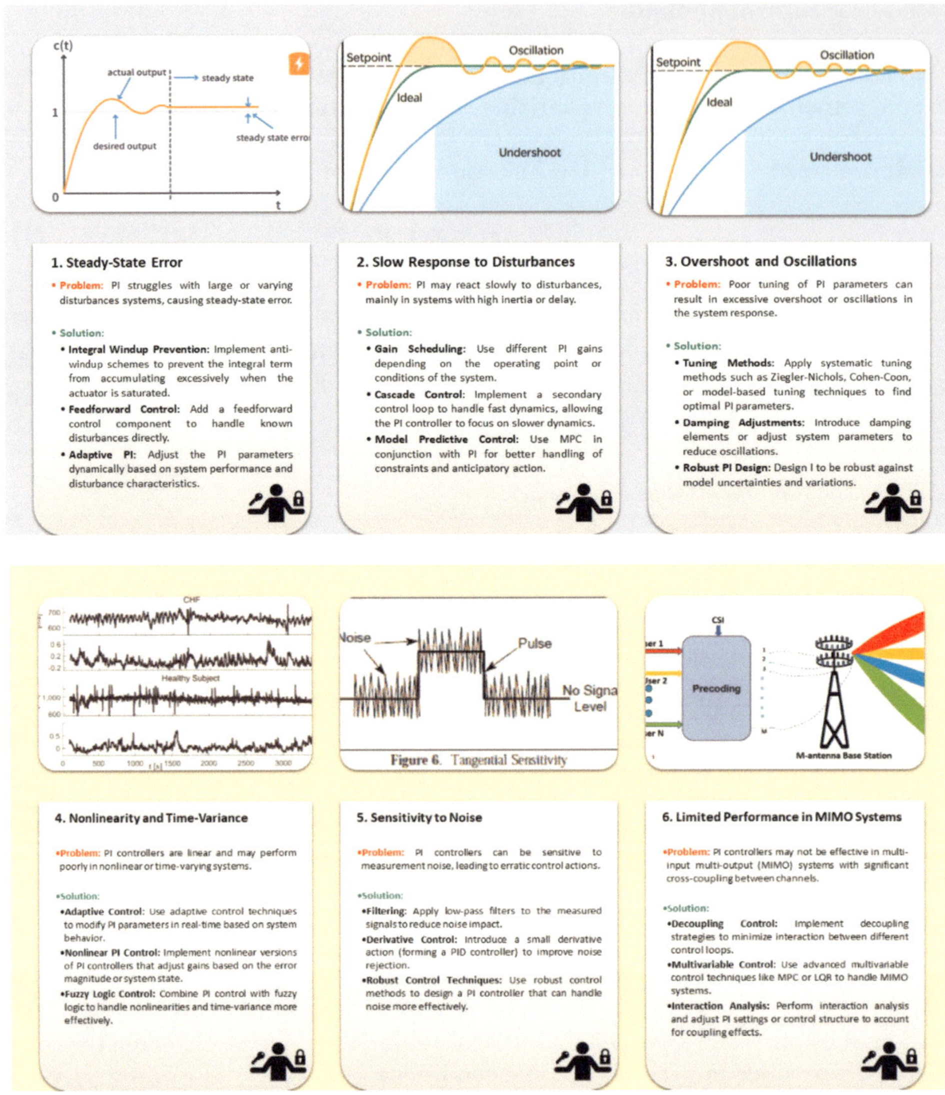

Figure 1.
The solutions to solve PI constraints and failures.

The $s(t)$ denotes the sliding variable, while λ and k are positive control gains that influence the convergence speed and robustness of the system.

1.5.1 Enhancements over traditional SMC

STSMC significantly enhances the performance and robustness of traditional SMC by addressing several key limitations [32, 44, 51, 78, 79].

- *Reduced chattering*: STSMC provides a continuous control input, reducing high-frequency chattering present in traditional SMC.

- *Improved smoothness*: Ensures smoother system responses, reducing mechanical wear and tear.
- *Better disturbance rejection*: Enhances the system's ability to handle disturbances and uncertainties.
- *Enhanced robustness*: Offers superior performance in systems with nonlinearities and uncertainties.
- *Precision in control*: Suitable for complex applications requiring high precision, such as renewable energy systems and electric drives.

1.5.2 Optimal control strategies

Table 2 provides a comparative overview of various optimal control methods, detailing their key features and characteristics including LQR [58, 80], LQR with Integral Action [78], Model Predictive Control (MPC) [81], Dynamic Programming [82], Pontryagin's Minimum Principle [83], H-infinity Control [84], and Optimal Stochastic Control [85]. This comparison helps understand each control strategy's strengths and limitations for different applications and system requirements.

2. Modeling of the system under study

The provided schematic in **Figure 2** illustrates the integration and control of various components within an energy management system (EMS) for a standalone DC-microgrid system (DC-MG). The system includes multiple energy sources and storage devices, all managed by a central control unit.

1. *Energy Sources:* The PV panels, diesel generator, and microgrid system are modeled as controlled current sources to reduce system simulation time, allowing a focused analysis of the contribution of HESS devices.

2.1 Photovoltaic (PV) panels

Represented as controlled current sources ($i_{PV}{}^{Ref}$) to simulate the PV output. Mathematical modeling of setpoint PV current includes the input variables: Irrad: Solar irradiance input, $V_{dc:}$ DC bus voltage, P_{pv}: Power from the PV panels, and the $P_{pv\_ref:}$: Reference power for the PV panels.

Below, Eq. (5) defines the PV setpoints current and power using regular MPPT operating mode. Instead, Eq. (6) defines the PV setpoints current under perturbed power balance cases using the FPPT operating mode to dampen the PV power generation. Finally, the PV system can switch between the two above modes regarding the system energy management scenarios, as noted in Eq. (7).

MPPT Mode: The goal is to maximize the PV power extraction:

$$\begin{cases} i_{PV_{-ref}}\ \mathrm{MPPT} = \dfrac{P_{\mathrm{MPPT}}}{V_{\mathrm{dc}}} \\ P_{MPPT} = \dfrac{\mathrm{Irrad}}{1000} \times 31527\ \text{(conversion factors based on the model)} \end{cases} \tag{5}$$

Control method	System type	Cost function	State feedback	Prediction horizon	Optimization type	Handling constraints	Robustness	Implementation complexity	Steady-state error handling	Computational cost	Real-time applicability	Uncertainty handling
LQR	Linear	Quadratic	Yes	Infinite	Algebraic Riccati Equation	No	Moderate	Moderate	No	Low	Yes	Limited
LQR with integral action	Linear with steady-state error	Quadratic with integral error	Yes	Infinite	Augmented Algebraic Riccati Equation	No	Moderate	Moderate	Yes	Low	Yes	Limited
MPC	Linear/ Nonlinear	Quadratic or custom	Yes	Finite	Quadratic Programming / Constrained Optimization	Yes	High	High	Yes	High	Challenging	Limited
Dynamic programming	Discrete/ Continuous	General	Yes	N/A	Recursive/ Iterative	Limited	Moderate	High	Possible	High	Challenging	Limited
Pontryagin's minimum principle	Continuous	General	Yes	N/A	Boundary Value Problem	Limited	Moderate	High	Possible	High	Challenging	Limited
H-infinity control	Linear/ Nonlinear	H-infinity norm	Yes	N/A	Frequency Domain	Yes	High	High	Yes	Moderate	Possible	High
Optimal stochastic control	Stochastic	Expected quadratic	Yes	N/A	Stochastic Differential Equations	Limited	Moderate	High	Yes	High	Possible	High

Table 2.
Comparison summarizing the key features of various optimal control strategies.

FPPT Mode: The EMS decides the setpoint P_{Pv-ref} to dampen the PV power.

$$i_{PV\_ref\_FPPT} = \frac{P_{Pr\_ref}}{V_{dc}} \tag{6}$$

Switching Modes: Selects MPPT or FPPT based on operational requirements:

$$i_{PV\_ref} = \begin{cases} i_{PV\_ref\_MPPT} & \text{if MPPT mode is ON} \\ i_{PV\_ref\_FPPT} & \text{if FPPT mode is ON} \end{cases} \tag{7}$$

2.2 Internal diesel generator (IDG)

Replaced with a controlled current source (I_{DG}^{Ref}) to simulate the generator output. Mathematical Modeling of Setpoint DG Current in Eq. (8-c) includes the input variables, which are the P_{DG_out} denoting the Power output of the Diesel Generator (DG) and the V_{DC}. Then, the output of the DG is the fuel consumption parameters for 48 kW DG including fuel curve intercept coefficients (a_0): 0.01458 L/hr/kW, fuel curve slope coefficient (a_1): 0.2333 L/hr/kW, and the nominal power (P_{nom}) : 48 kW. Thus, the setpoint current i_{DG} ref. for the DG is:

$$\begin{cases} F(P_{DG\_out}) = a_0 + a_1 \cdot P_{DG\_out} + a_2 \cdot P_{DG\_out}^2 \\ F(P_{DG\_out}) = 0.01458 + 0.2333 \cdot P_{DG\_out} \\ i_{DG\_ref} = \dfrac{P_{DG\_out}}{V_{dc}} \end{cases} \tag{8}$$

2.3 Nearby DC-microgrids (MGs)

Simulated using a controlled current source (iMGS_ref) to represent the nearby DC microgrids.

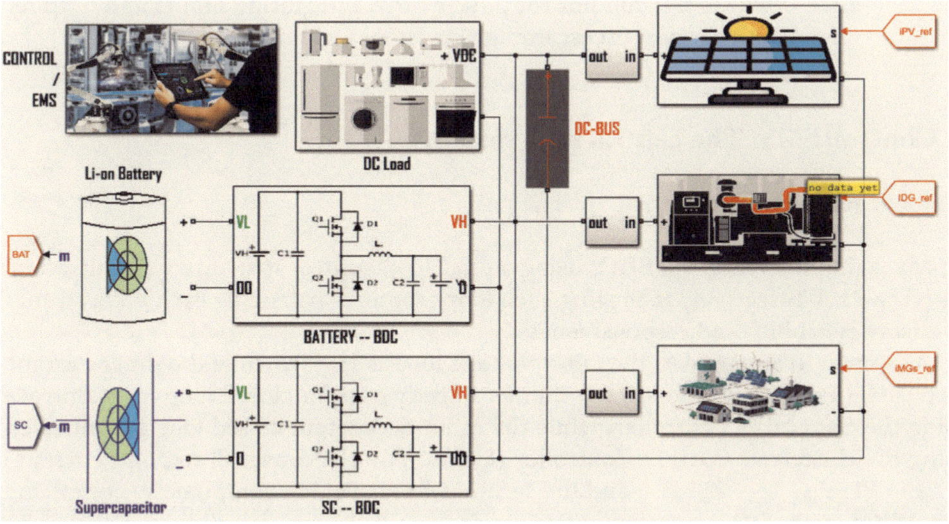

Figure 2.
The proposed standalone DC-MG under study.

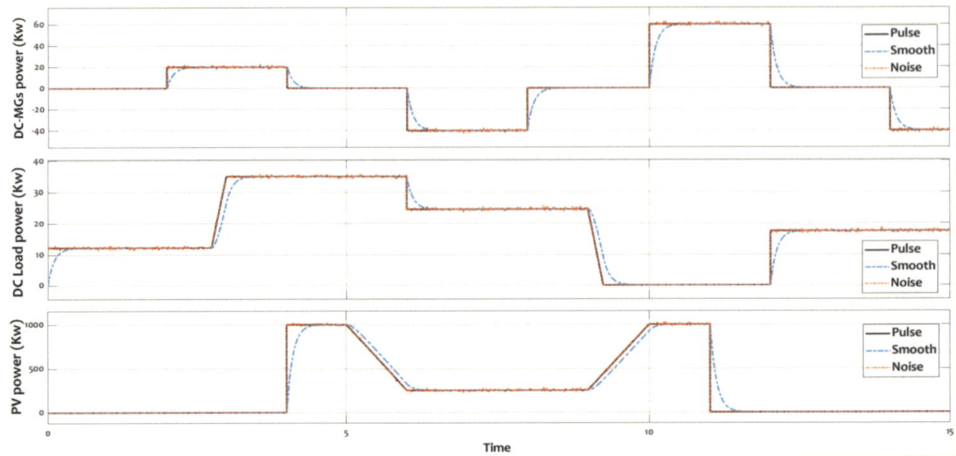

Figure 3.
The proposed waveform of the system resources under study.

3. Energy storage systems (ESS)

a. *Li-ion Battery:* Managed by a bidirectional DC-DC converter (BDC), controlling low (VL) and high (VH) voltage levels and during charge-discharge cycles.

b. *Supercapacitor (SC):* Also managed by a BDC, for rapid charge/discharge cycles, controlling low (VL) and high (VH) voltage levels.

1. *DC Load:* Includes a dynamic DC load with three profiles (smooth, pulsing, and perturbing waveforms) as shown in **Figure 3**. These profiles are proposed for the performance assessment of the studied system.

2. *DC-Bus:* Central conduit for power flow, integrating inputs and outputs from all energy sources, storage systems, and loads.

4. Control/EMS: The central unit responsible for

4.1 Charge-discharge control

Managing the HESS via BDCs using a dual-loop control structure with filtration-based control. Effectively managing such loops requires balancing each ESS's strengths to ensure reliability and responsiveness.

As shown in **Figure 4A**, the outer voltage loop is $H_V(s)$, while the inner current loop is $H_C(s)$. The outer voltage and inner battery current closed loops are controlled using the classical PI controller, while the inner SC current closed loop is controlled using the Hysteresis Current Controller (HCC). The HESS overall setpoint current is $I_{L\_ref} = \text{PI}\ (V_{dc\_ref} - V_{dc\_mes})$, and the setpoint duty cycle to compute the complementary PWM to the BDC switches is $Duty\ cycle_{ref} = \text{PI}\ (i_{L\_ref} - i_L)$. [24, 25, 88].

In **Figure 4B**, the total current is then split into $i_{B\ ref}$ and $i_{SC\ ref}$ reference currents using a rate limiter and an LPF. The advanced LPF (ALPF) incorporates a

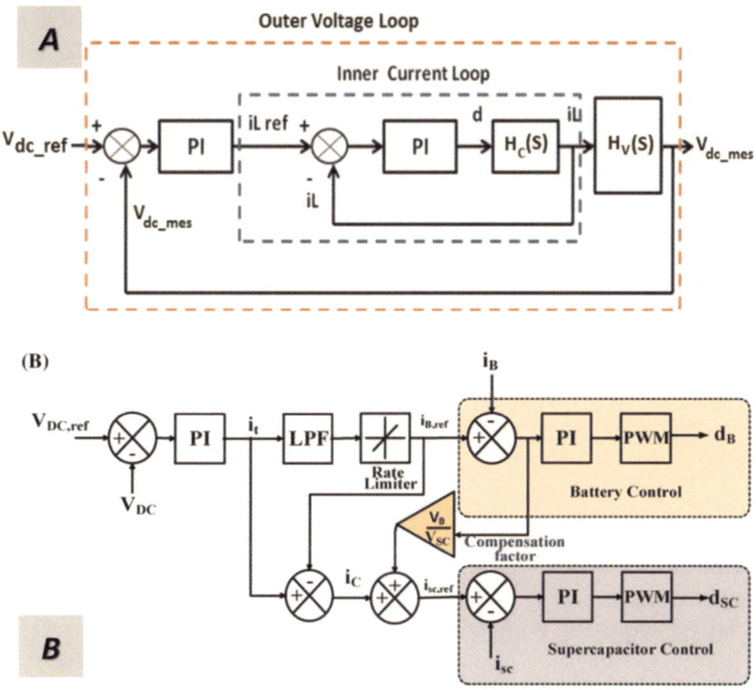

Figure 4.
Hierarchical control: A- Dual loop control of Single ESS, B- ALPF control of HESS [86, 87].

compensation factor of $C = \frac{V_b}{Vsc}$ to optimize the reference split between the battery and SC, improving DC voltage regulation.

The total 'it' is adjusted, with the battery reference $i_{B,ref} = rate\ limiter\ (LPF(it)) \times C$ and the SC reference $i_{SC\ ref} = i_t - i_{B\ ref}$ [24, 25, 88].

4.2 Energy management

Governing power balance among DERs based on solar irradiation, diesel fuel availability, HESS State Of Charge (SOC), load demand, and the power balance status of nearby DC-MGs to share excess/lack of energy.

The EMS optimizes power allocation by coordinating inputs from the PV, IDG, and MGs (controlled current sources). This approach reduces simulation time and focuses on the control and HESS-EMS through their DC-DC BDCs. The dual-loop control structure with filtration-based control ensures efficient charge/discharge management, while the EMS maintains power balance and shares energy among resources, ensuring system stability and efficiency.

a. *Initialization:* Initializes system parameters and states, ensuring consistent starting conditions for the EMS function: P_{PV}^{rated} = 50e3; $SOCB_{min}$ = $SOCsc_{min}$ = 0.2; $SOCB_{max}$ = $SOCsc_{max}$ = 0.8; I_{MGs_ref} = I_{DG}^{Ref} = P_{PV}^{Ref} = state = 0; S^{LOAD} = mppt = DG = 1; i_{PV}^{Ref} = (Irrad * P_{PV}^{rated}) / Vdc * 1e3)

b. *Power Balance Calculation:* Computes local (LNP) and global (GNP) power balances to assess the current energy demand and supply status.

c. *SOC Intervals:* Determines the working state of the battery and SC based on their SOC, categorizing them into three states: low, normal, and high. This helps in making informed decisions on charging or discharging.

If SOCB < $SOCB_{min}$;	$SOCB_{status}$ = 3
Elseif SOCB > = $SOCB_{min}$ and SOCB <= $SOCB_{max}$;	$SOCB_{status}$ = 1
Elseif SOCB > $SOCB_{max}$;	$SOCB_{status}$ = 2

a. *Power Balance:* Each scenario is handled by adjusting the reference currents (I_{DG}^{Ref} and I_{MG}^{Ref}) and states to maintain system stability and efficiency. Manages different scenarios of LNP and GNP values:

- Stable power balance with no energy demand.
- Excess power is available to be absorbed by the system.
- Power deficit requiring energy export or import.

a. *Outputs:* Outputs the references [I_{DG}^{Ref}, I_{MG}^{Ref}, state, S^{LOAD}, P_{PV}^{Ref}, i_{PV}^{Ref}] necessary for controlling the system, ensuring the EMS can implement the calculated strategies effectively.

Table 3 provides a clear understanding of the EMS, emphasizing the SOC-EMS importance, power balance scenarios, and coordinated HESS control and nearby MGs.

5. Simulation results with discussion

In this section, the studied standalone system is simulated using MATLAB Simulink, employing three main topologies with several scenarios for performance assessment under various constraints of system resources, loads, and control settings. The main parameters of the studied energy system are classified in **Table 4**.

The sections below provide analyses/discussions of the revealed results.

5.1 System power balance and DC bus voltage stability

Figure 5 illustrates the system power balance and DC bus voltage using PI controllers under three different waveform conditions: pulse, smooth, and perturbed.

For pulse waveforms (A), the power balance shows significant fluctuations that correspond to the pulsed load. The DC bus voltage maintains regulation but exhibits noticeable ripples due to the abrupt changes. This indicates that the PI can handle pulse loads but with some difficulty in maintaining a smooth voltage profile.

In the case of smooth waveforms (B), the power balance is more stable compared to the pulse waveforms. The DC bus voltage is well-regulated with minimal ripples, indicating better performance of the PI in steady-state conditions where the load changes are gradual and predictable.

Condition	SOC $_{Bat}$	DG^{status}	$I_{DG}{}^{Ref}$	$I_{MG}{}^{Ref}$	State
LNP = GNP =0	Any	Any	0	0	0.5
LNP = 0 & GNP > 0	1 or 3		0	GNP/V_{DC}	1
	2		0	0	1.5
LNP = 0 & GNP < 0	1 or 3	0	0	0	2
		1	−GNP/V_{DC}	GNP/V_{DC}	2.5
LNP = 0 & GNP < 0	2	0	0	GNP/V_{DC}	3
		1	−GNP/V_{DC}	GNP/V_{DC}	3.5
LNP > 0 & GNP = 0	1 or 2	0	0	0	4
		1	0	0	4.5
LNP > 0 & GNP = 0	3	0	0	0	5
		1	P_{Load}/V_{DC}	0	5.5
LNP > 0 & GNP > 0 • GNP > = LNP	1 or 3	Any	0	GNP/V_{DC}	6
	2		0	GNP/V_{DC}	6.5
LNP > GNP	1 or 2		0	GNP/V_{DC}	7
LNP > GNP	3	0	0	GNP/V_{DC}	7.5
		1	$\frac{\lvert GNP\rvert-\lvert LNP\rvert}{V_{DC}}$	GNP/V_{DC}	8
LNP > 0 & GNP < 0	1 or 2	0	0	0	8.5
	3		0	0	9
	1 or 2	1	−GNP/V_{DC}	GNP/V_{DC}	9.5
	3		LNP/V_{DC}	0	10
LNP < 0 & GNP = 0	1 or 3	Any	0	0	10.5
	2		0	0	11
LNP < 0 & GNP > 0	1 or 3		0	GNP/V_{DC}	11.5
	2		0	0	12
	1 or 2	$\lvert LNP\rvert \geq \lvert GNP\rvert$	0	GNP/V_{DC}	12.5
		$\lvert GNP\rvert < \lvert LNP\rvert$ & DG = 0	0	GNP/V_{DC}	13
		$\lvert GNP\rvert < \lvert LNP\rvert$ & DG = 1	$\frac{\lvert GNP\rvert-\lvert LNP\rvert}{V_{DC}}$	GNP/V_{DC}	13.5
	3	0	0	0	14
		1	−GNP/V_{DC}	GNP/V_{DC}	14.5

Table 3.
Energy management strategy rules based on system resources and constraints.

For perturbed waveforms (C), the power balance fluctuates significantly due to the perturbations in the load. The DC bus voltage exhibits more significant waves, indicating that the PI controller has difficulty sustaining precise voltage regulation during dynamic situations. This underscores the limitations of the PI in contexts characterized by swift load fluctuations.

• PV: Peak power: 50 (Kw); DG: Peak power: 48 (Kw); • Nearby DC-MGs: Peak power ± 15 (Kw);	*PV/DG/Nearby DC-MGs*
• Cut-off, nominal, and fully charged voltage: 123–165-192 (V) • Rated, maximum, and nominal voltage capacity: 10–10 – 0.49 (Ah) • Nominal discharge current: 4.3478 (A); • Internal resistance: 0.165 (Ω)	Lithium-ion battery
• Rated voltage / Rated capacitance: 167.4 (V) / 2(F); • Equivalent DC series resistance: 1.4 (Ω)	SC bank
• Rated DC Bus voltage: 250 (V) / Peak power: 35 (Kw); • DC Bus capacitance: 100 (uF)	DC variable Load
• Rated input voltage/Rated output voltages: 165 / 250 (V) • Inductance: 1(mH); Capacitance: 0.05(F); • Switching frequency: 20(kHz);	DC/DC Bidirectional Converter

Table 4.
Overview of reel-implemented DC-MGs around the world.

Figure 6 illustrates the system power equilibrium and DC bus voltage utilizing the ST-SMC over the identical three waveforms. According to the pulse waveforms (A), there are less swings in power balance in comparison with the power balance (PI), which indicates that the control of rapid load variations gets better. In dynamic settings, the ST-SMC regulates DC bus voltage more effectively and with fewer ripples, proving its resilience. This shows that the ST-SMC performs better under pulse loads.

Smooth waveforms (B) have a constant power balance, similar to the PI but with fewer changes. Due to its systematic DC bus voltage maintenance with little ripples, the ST-SMC has excellent steady-state performance. This suggests that ST-SMC can maintain a more stable voltage profile under steady load.

When perturbed waveforms (C) are evaluated, the power balance varies less than the PI, indicating better disturbance rejection. The capacity of the ST-SMC to maintain voltage stability in the face of disturbances is demonstrated by the fact that the DC bus voltage is successfully regulated with minimal ripples. Based on this, it can be concluded that ST-SMC is more effective in environments that are defined by rapid and unpredictable swings in load.

5.2 HESS results

Figure 7 shows the HESS performance with the PI controllers for SOC, duty cycle, and voltages.

- *Battery SOC*: The PI controller exhibits minor fluctuations in the battery SOC across all wave conditions, starting around 0.50 and gradually decreasing to around 0.45 by the end of the observation period. In comparison, the ST-SMC follows a similar decreasing trend but shows slightly more stability, especially under perturbed conditions. The final SOC value is ≈0.45, indicating similar overall energy usage but potentially better management of transitional fluctuations by the ST-SMC.

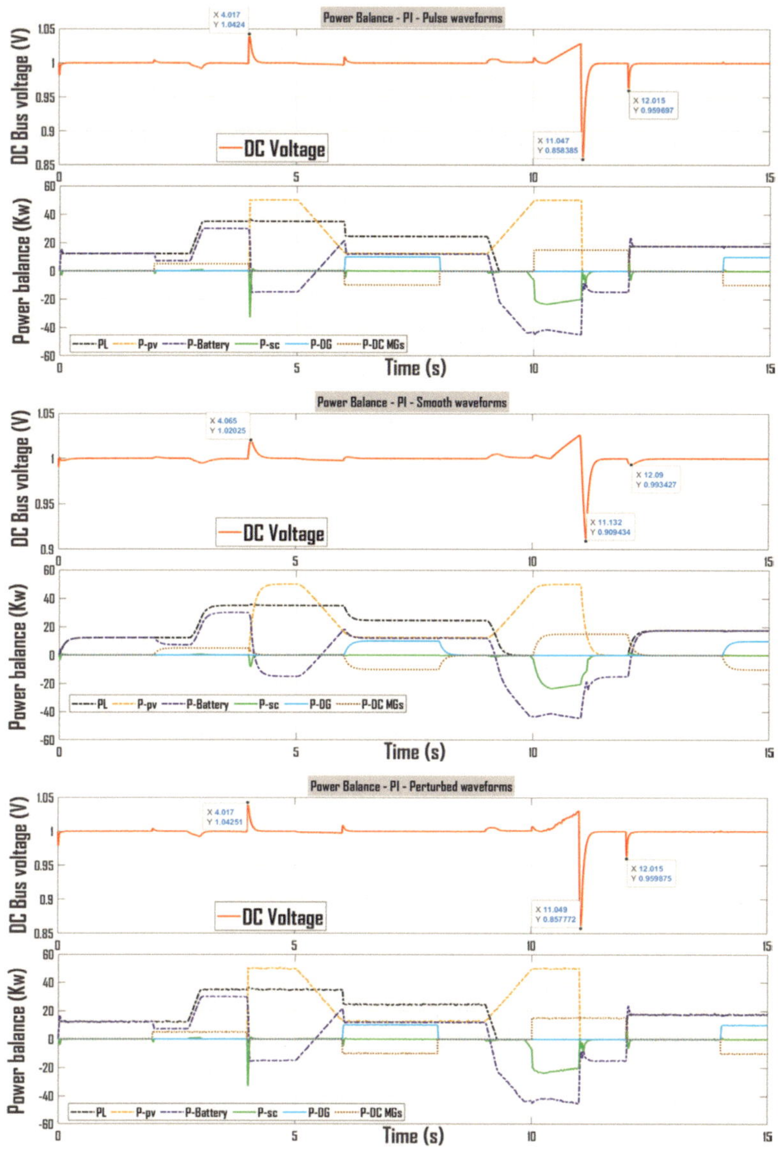

Figure 5.
System power balance and DC bus voltage with PI controllers: A - Pulse, B - Smooth, C - Perturbed waveforms.

- *Battery Duty Cycle*: The PI controller's duty cycle fluctuates significantly under all wave conditions, particularly under perturbed conditions where the fluctuations are more pronounced, varying between 0.2 and 0.8. In contrast, the ST-SMC controller exhibits a more controlled duty cycle response, with less extreme fluctuations, suggesting more efficient handling of load changes.

- *Battery Voltage*: The PI exhibits clear fluctuations in battery voltage, reflecting variations in load conditions, with values between 0.7 and 1.4. The voltage fluctuations are sudden, especially in disturbed situations. The ST-SMC exhibits more gradual voltage shifts, maintaining a comparable voltage range of 0.75 to

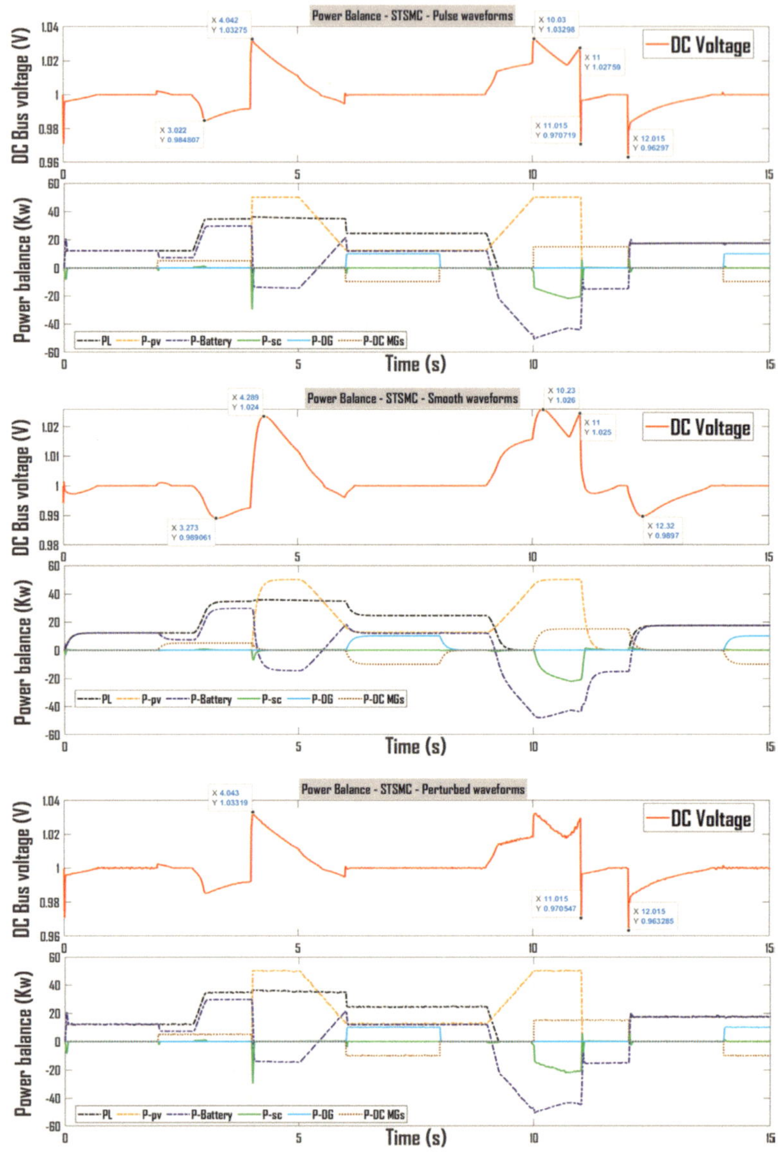

Figure 6.
System power balance and DC bus voltage with ST-SMC: A - Pulse, B - Smooth, C - Perturbed waveforms.

1.35, which signifies enhanced stability and improved management of load variations.

Figure 8 illustrates the HESS performance utilizing the STSMC controllers for the state of charge, duty cycle, and voltages.

- *SC SOC:* The state of charge (SOC) of the supercapacitor in the PI controller exhibits considerable changes, particularly under disturbed conditions, oscillating swiftly between 0.5 and 0.8 with prominent spikes. The ST-SMC controller offers enhanced stability for the supercapacitor's state of charge,

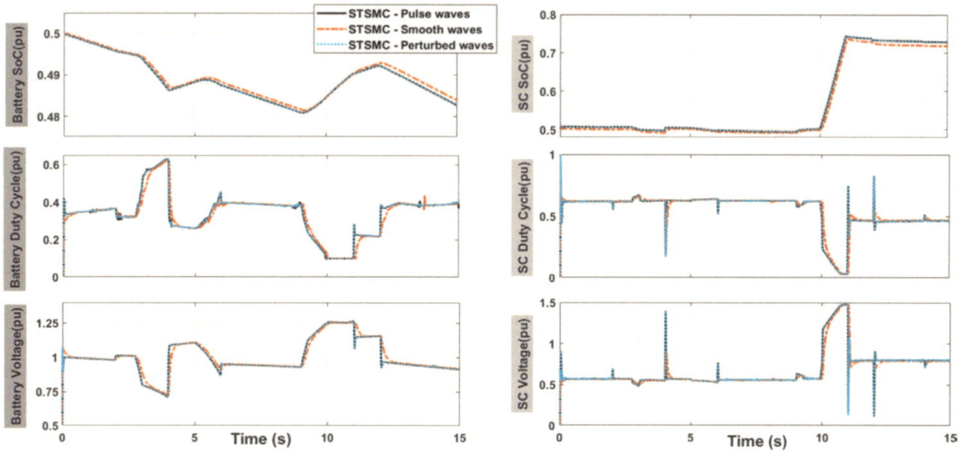

Figure 7.
HESS curves with PI controller: A - SOC, B - duty cycle, and C - Voltages.

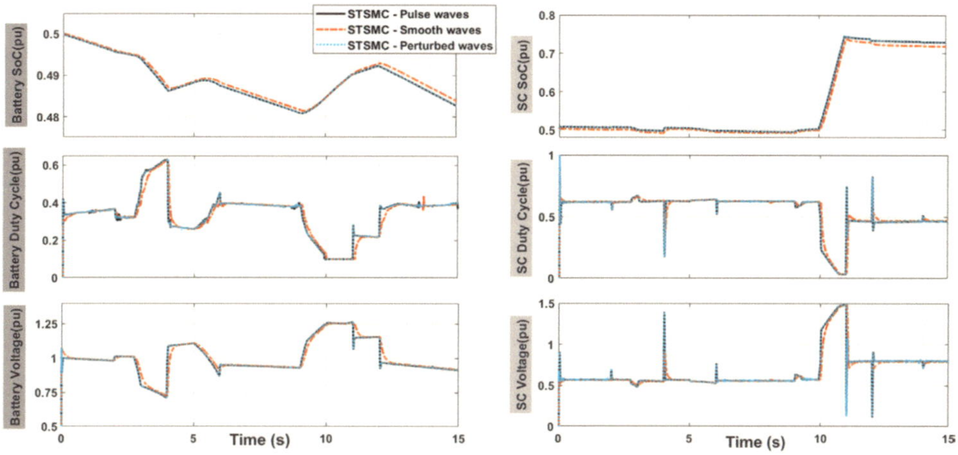

Figure 8.
HESS curves with ST-SMC controller: A - SOC, B - Duty cycle, and C - Voltages.

exhibiting a more gradual reaction to load variations and reduced spikes, signifying superior control and a more equitable energy distribution.

- *SC Duty Cycle:* The PI controller's supercapacitor duty cycle fluctuates between 0.2 and 0.8, especially in disturbed conditions, indicating a very sensitive control system. ST-SMC controllers have fewer unexpected oscillations due to their regular duty cycles. This suggests that the controller can improve energy distribution management and reduce system component wear.

- *SC Voltage:* The PI controller's supercapacitor voltage fluctuates abruptly when parameters are changed. These fluctuations range from 0.5 to 1.5, and voltage spikes imply reactive control. However, the ST-SMC provides smoother voltage shifts for the supercapacitor, which enhances system reliability and efficiency.

The comparative analysis found that the ST-SMC controller has better stability and control than the PI controller. Smoother transitions and fewer abrupt parameter variations distinguish it. This stability may improve system reliability, component stress, and lifespan.

5.3 EMS transition and performance

Figure 9 shows how PI and ST-SMC controllers transition between Energy Management System (EMS) rules and situations. EMS regulation transitions are smoother and more consistent with the ST-SMC controller than with the PI controller. The ST-SMC controller reduces oscillations and speed stabilization during EMS instance changes. This shows that the ST-SMC performs better during system state transitions, which is crucial for system stability. The results are fully analyzed and discussed in this section, with a focus on comparing PI with ST-SMC for performance evaluation using RMSE-MAE metrics. In the following text, figures show the outcome.

5.4 Analysis of RMSE and MAE for DC bus voltage (DV)

Figure 10 compares the PI and ST-SMC controllers' DC Bus voltage error (DV) RMSE and MAE metrics in various circumstances. The PI controller has an RMSE of 2.48 and an MAE of 0.61 under pulse conditions. The ST-SMC controller has a higher RMSE of 2.59 and MAE of 1.52. Considering this, the PI controller appears to minimize dynamic reaction errors during pulses. PI has a reduced RMSE of 1.86 and MAE of 0.54 under ideal conditions. In contrast, ST-SMC has an RMSE of 2.18 and an MAE of 1.28. The pattern persists when settings are modified, with the PI having a lower RMSE (2.50) and MAE (0.63) than the ST-SMC (2.59), which had a higher MAE (1.53). The PI reduces dynamic errors exceptionally well under all settings.

5.5 Analysis for battery current (Ib)

Figure 11 shows that the PI controller always reports much greater battery current error (DiLB) measurements. In the pulse scenario, the PI controller's RMSE is 184.23

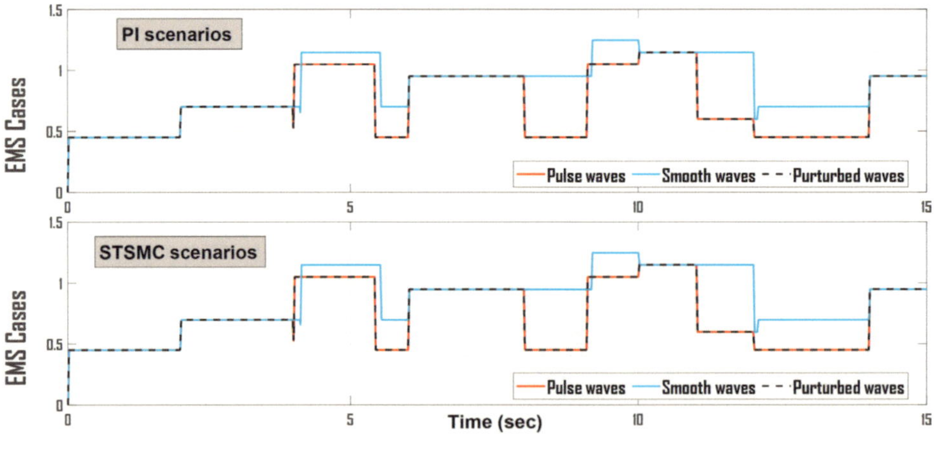

Figure 9.
The transition between EMS rules and cases using PI and ST-SMC controllers.

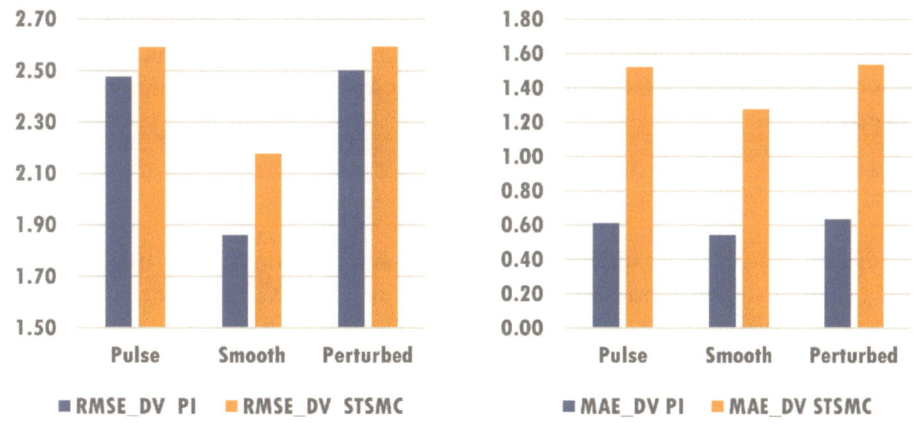

Figure 10.
The KPI charts of the DC bus voltage error.

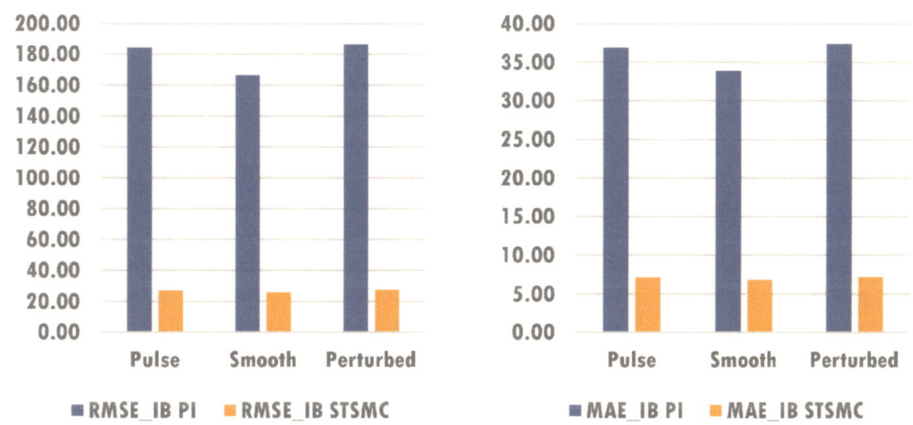

Figure 11.
The KPI charts of the battery current error.

and MAE is 36.91. ST-SMC controller values are much lower, with an RMSE of 27.13 and MAE of 7.10. This shows that the ST-SMC controller regulates battery variables better, reducing error. Under ideal conditions, the PI controller has an RMSE of 166.53 and an MAE of 33.88. The ST-SMC controller has significantly lower RMSE and MAE values of 25.96 and 6.78, respectively.

The perturbed PI has an RMSE of 186.41 and an MAE of 37.36. However, the ST-SMC controller performs better with an RMSE of 27.36 and MAE of 7.10. This constant performance shows that the ST-SMC controller controls battery variables more efficiently than the PI controller.

5.6 Analysis for Supercapacitor current (Isc)

The perturbed PI has an RMSE of 186.41 and an MAE of 37.36. However, the ST-SMC controller performs better with an RMSE of 27.36 and MAE of 7.10. This constant performance shows that the ST-SMC controller controls battery variables more efficiently than the PI controller. This pattern repeats under smooth conditions. The PI controller has an RMSE of 124.61 and an MAE of 23.95, while the ST-SMC controller has 4.72 and 0.65. The PI has an RMSE of 141.41 and an MAE of 25.31 when circumstances are disrupted, while the ST-SMC has 7.05 and 1.02. The ST-SMC controller outperforms the PI controller in overload current accuracy and control (**Figure 12**).

5.7 Analysis for load mismatch (DPLoad)

In **Figure 13**, load mismatch error measurements (DPLoad) show that both controllers function similarly under pulse settings. The ST-SMC controller has an RMSE of 0.59 and an MAE of 0.08, while the PI controller has 0.55 and 0.12.

Under optimal conditions, the PI controller has an RMSE of 0.27 and an MAE of 0.06, whereas the ST-SMC has 0.28 and 0.07. Both controllers have comparable RMSE values of 0.59 under perturbed circumstances, but the ST-SMC controller has a lower MAE of 0.07 than the PI's 0.08.

Although both controllers perform similarly in power loss, the ST-SMC reduces absolute errors better.

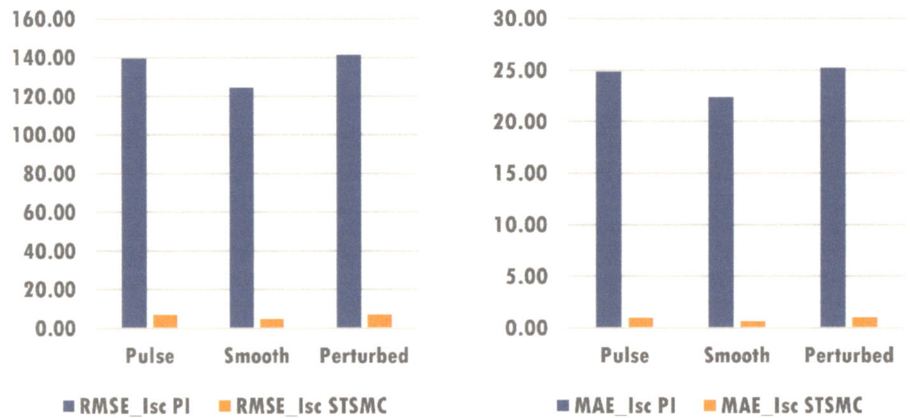

Figure 12.
The KPI charts of the SC current error.

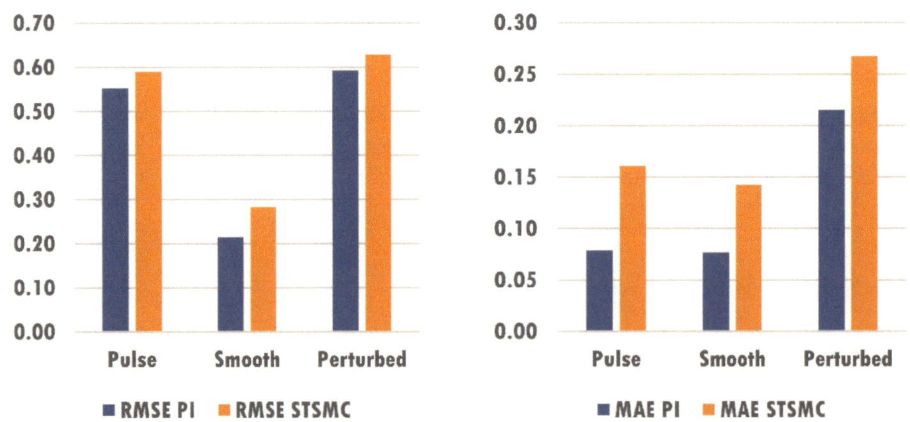

Figure 13.
The KPI charts of the load power mismatch (DPLoad).

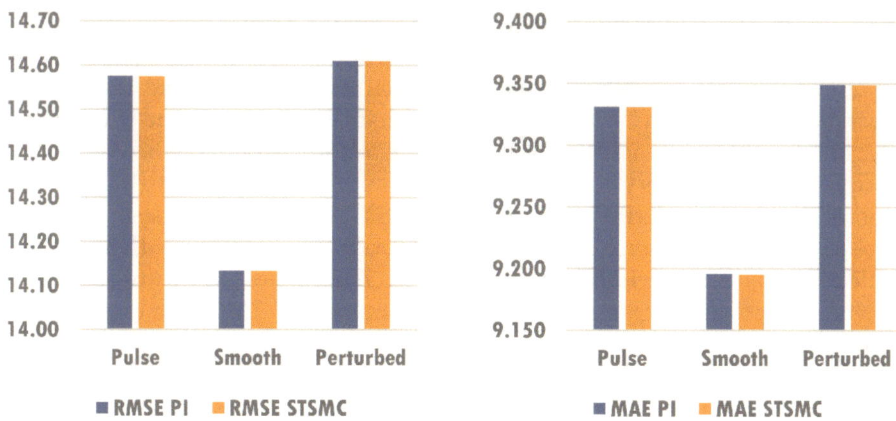

Figure 14.
The KPI charts of the total system power losses (DPLoss).

5.8 Analysis for system power losses (DPLoss)

Figure 14 shows that the PI and ST-SMC controllers have extremely similar RMSE and MAE values for system power losses (DPLoss). Under pulse conditions, both controllers have RMSE 14.58 and MAE 9.331. In optimal conditions, both controllers have RMSEs of 14.57 and MAEs of 9.332. Both controllers' RMSE approximates 14.61 under perturbed conditions, whereas MAE equals 9.349. The two controllers perform similarly in reducing load divergence from the setpoint.

A comprehensive investigation of RMSE and MAE metrics across system variables indicated that the PI and ST-SMC perform differently. DC Bus voltage (DV) errors are generally reduced by the PI, lowering RMSE and MAE. ST-SMC has better precision and regulation than PI for battery current (IB) and short-circuit current. Load mismatch and DPLoad work similarly, but ST-SMC reduces absolute errors better. Total system power losses (DPLoss) are similar for both. PI thrives in dynamic circumstances, but ST-SMC improves battery management and SC current tracking control and stability, according to the study. For systems that need precise energy control and stability, the ST-SMC is more trustworthy.

6. Conclusion

An independent energy system using PI and ST-SMC was simulated to evaluate system power balance, DC bus voltage stability, and EMS transitions. These results were tested under various waveforms. The whole investigation showed that various components regulated system variables differently.

Pulsed conditions increased the PI's DC Bus voltage error (DV) maintenance, with an RMSE of 2.48 and an MAE of 0.61. PI readings decreased similarly under calm and disturbed situations. Despite this, the ST-SMC regulated battery and supercapacitor currents better. Under pulse conditions, the PI controller had an RMSE of 184.23 and an MAE of 36.91 for battery current. The ST-SMC has a substantially lower RMSE of 27.13 and MAE of 7.10 for the same battery current. Its continuous trend across configurations showed that the ST-SMC increased battery management precision and robustness.

The ST-SMC excelled at power distribution and load fluctuations. Both techniques performed similarly under load divergence or mismatch (DPLoad); however, the ST-SMC reduced absolute errors better. Both exhibited similar RMSE and MAE values for total system power losses (DPLoss), indicating that they reduced load divergence equally. In conclusion, the PI minimizes dynamic response faults, whereas the ST-SMC provides better battery and supercapacitor control and stability. It has smoother transitions and slower volatility than the PI. The ST-SMC is ideal for systems that need accurate energy management and stability, especially in situations with quick and unpredictable load fluctuations, the study found. The ST-SMC may improve system dependability, component stress, and lifespan. Because ST-SMC provides more stability and control.

Acknowledgements

We extend our deepest gratitude to the authors and researchers whose groundbreaking work has significantly contributed to the success of this book chapter. Their insightful studies and innovative theories have provided a robust foundation upon which this chapter is built. We appreciate their dedication to advancing knowledge and their willingness to share their intellectual contributions with the broader academic community.

Conflict of interest

The authors declare that there is no conflict of interest regarding the publication of this book chapter. All research and findings presented are the result of impartial and objective analysis, conducted without any influence from personal, financial, or professional interests.

Acronym and abbreviation

AC	alternating current
AI	artificial intelligence
ALPF	advanced low pass filter
ANFIS	adaptive neuro-fuzzy inference system
BDC	bidirectional DC-DC converter
DC	direct current
DERs	distributed energy resources
DG	diesel generator
DFIG	doubly fed induction generator
DV	DC bus voltage error
EMS	energy management system
ESS	energy storage systems
EU	European Union
EV	electric vehicle
FLC	fuzzy logic controller
FOPID	fractional order proportional-integral-derivative
FPPT	forced power point tracking

DOI: http://dx.doi.org/10.5772/intechopen.1006834

GA	genetic algorithm
GNP	global net power
HCC	hysteresis current controller
HESS	hybrid energy storage system
HVDC	high voltage direct current
HVAC	high voltage alternating current
IDG	setpoint current of diesel generator
LNP	local net power
LPF	low pass filter
LQR	linear quadratic regulator
LVAC	low voltage alternating current
MAE	mean absolute error
MG	microgrid
MPC	model predictive control
MPPT	maximum power point tracking
MVAC	medium voltage alternating current
PEM	proton exchange membrane
PI	proportional-integral
PID	proportional-integral-derivative
PMSM	permanent magnet synchronous motor
PV	photovoltaic
PWM	pulse width modulation
RE	renewable energy
RES	renewable energy sources
RMSE	root mean square error
SC	supercapacitor
SMC	sliding mode control
SOC	state of charge
ST-SMC	super twisting sliding mode control
U.S DOE	United States Department of Energy
UUV	unmanned underwater vehicle
VDC	DC bus voltage
WT	wind turbine

Author details

Mohamed Amine Hartani[1,2*], Aissa Benhammou[2] and Abdallah Laidi[1]

1 University Ahmad Draia - LDDI, Adrar, Algeria

2 University Tahri Mohamed - SGRE, Bechar, Algeria

*Address all correspondence to: ami.hartani@univ-adrar.edu.dz

References

[1] Hunjra AI, Hassan MK, Zaied YB, Managi S. Nexus between green finance, environmental degradation, and sustainable development: Evidence from developing countries. Resources Policy. 2023;**81**:103371

[2] Xie Y, Zhang Y, Lee W-J, Lin Z, Shamash YA. Virtual power plants for grid resilience: A concise overview of research and applications. IEEE/CAA Journal of Automatica Sinica. 2024;**11**: 329-343

[3] Sani SB, Celvakumaran P, Ramachandaramurthy VK, Walker S, Alrazi B, Ying YJ, et al. Energy storage system policies: Way forward and opportunities for emerging economies. Journal of Energy Storage. 2020;**32**: 101902

[4] Fotopoulou M, Pediaditis P, Skopetou N, Rakopoulos D, Christopoulos S, Kartalidis A. A review of the energy storage systems of non-interconnected European Islands. Sustainability. 2024;**16**:1572

[5] Hamilton J, Negnevitsky M, Wang X, Lyden S. High penetration renewable generation within Australian isolated and remote power systems. Energy. 2019;**168**:684-692

[6] Ramanan C, Roy S, Yam KS, Lim KH, Bora BJ, Medhi BJ. Floating solar PV systems—Global research reported in the year 2022. In: Singh SDKM, Subudhi S, Singh NK, editors. Conference on Fluid Mechanics and Fluid Power. Roorkee, India: Springer; 2022. pp. 61-78

[7] Supapo KRM, Lozano L, Tabañag IDF, Querikiol EM. A backcasting analysis toward a 100% renewable energy transition by 2040 for off-grid islands. Energies. 2022;**15**:4794

[8] Pinto G, Naspolini H, Rüther R. Assessing the economic viability of BESS in distributed PV generation on public buildings in Brazil: A 2030 outlook. Renewable Energy. 2024;**225**:120252

[9] Oyarzún-Aravena AM, Chen J, Brownbridge G, Akroyd J, Kraft M. An Analysis of Renewable Energy Resources and Options for the Energy Transition in Chile. Available from: SSRN 4906098

[10] Jing W, Lai CH, Wong WS, Wong MD. A comprehensive study of battery-supercapacitor hybrid energy storage system for standalone PV power system in rural electrification. Applied Energy. 2018;**224**:340-356

[11] Basu M. Dynamic optimal power flow for isolated microgrid incorporating renewable energy sources. Energy. 2023; **264**:126065

[12] Fathy A, Ferahtia S, Rezk H, Yousri D, Abdelkareem MA, Olabi A. Optimal adaptive fuzzy management strategy for fuel cell-based DC microgrid. Energy. 2022;**247**:123447

[13] Mengelkamp E, Gärttner J, Rock K, Kessler S, Orsini L, Weinhardt C. Designing microgrid energy markets: A case study: The Brooklyn microgrid. Applied Energy. 2018;**210**:870-880

[14] Cao W, Xiao J-W, Cui S-C, Liu X-K. An efficient and economical storage and energy sharing model for multiple multi-energy microgrids. Energy. 2022;**244**: 123124

[15] Gargari MZ, Hagh MT, Zadeh SG. Preventive scheduling of a multi-energy microgrid with mobile energy storage to enhance the resiliency of the system. Energy. 2023;**263**:125597

[16] Rangel N, Li H, Aristidou P. An optimisation tool for minimising fuel consumption, costs and emissions from diesel-PV-battery hybrid microgrids. Applied Energy. 2023;**335**:120748

[17] Adefarati T, Bansal R, Bettayeb M, Naidoo R. Optimal energy management of a PV-WTG-BSS-DG microgrid system. Energy. 2021;**217**:119358

[18] Guichi A, Mekhilef S, Berkouk E, Talha A. Optimal control of grid-connected microgrid PV-based source under partially shaded conditions. Energy. 2021;**230**:120649

[19] Reddy Y, Jithendranath J, Chakraborty AK, Guerrero JM. Stochastic optimal power flow in islanded DC microgrids with correlated load and solar PV uncertainties. Applied Energy. 2022;**307**:118090

[20] Jithendranath J, Das D, Guerrero JM. Probabilistic optimal power flow in islanded microgrids with load, wind and solar uncertainties including intermittent generation spatial correlation. Energy. 2021;**222**:119847

[21] Restrepo M, Cañizares CA, Simpson-Porco JW, Su P, Taruc J. Optimization- and rule-based energy management systems at the canadian renewable energy laboratory microgrid facility. Applied Energy. 2021;**290**: 116760

[22] Wu J, Yang F. A dual-driven predictive control for photovoltaic-diesel microgrid secondary frequency regulation. Applied Energy. 2023;**334**: 120652

[23] Qiu H, You FJAE. Decentralized-distributed robust electric power scheduling for multi-microgrid systems. Applied Energy. 2020;**269**: 115146

[24] Amine HM, Aissa B, Rezk H, Messaoud H, Othmane A, Saad M, et al. Enhancing hybrid energy storage systems with advanced low-pass filtration and frequency decoupling for optimal power allocation and reliability of cluster of DC-microgrids. Energy. 2023;**282**:128310

[25] Hartani MA, Rezk H, Benhammou A, Hamouda M, Abdelkhalek O, Mekhilef S, et al. Proposed frequency decoupling-based fuzzy logic control for power allocation and state-of-charge recovery of hybrid energy storage systems adopting multi-level energy management for multi-DC-microgrids. Energy. 2023;**278**: 127703

[26] Ferahtia S, Djeroui A, Rezk H, Houari A, Zeghlache S, Machmoum M. Optimal control and implementation of energy management strategy for a DC microgrid. Energy. 2022;**238**:121777

[27] Ghanbari A, Karimi H, Jadid S. Optimal planning and operation of multi-carrier networked microgrids considering multi-energy hubs in distribution networks. Energy. 2020; **204**:117936

[28] Abdelsalam M, Diab HY, El-Bary A. A metaheuristic Harris hawk optimization approach for coordinated control of energy management in distributed generation based Microgrids. Applied Sciences. 2021;**11**:4085

[29] Nallolla CA, Chittathuru VPD, Padmanaban S. Multi-objective optimization algorithms for a hybrid AC/DC microgrid using RES: A comprehensive review. Electronics. 2023;**12**:1062

[30] Jain D, Saxena D. Comprehensive review on control schemes and stability investigation of hybrid AC-DC

microgrid. Electric Power Systems Research. 2023;**218**:109182

[31] Astriani Y, Fauziah K, Hilal H, Prasetyo B. Load sharing control between PV power plant and diesel generator to mitigate effect of PV fluctuation using PID algorithm. In: 2017 International Conference on High Voltage Engineering and Power Systems (ICHVEPS). IEEE; 2017. pp. 140-144

[32] Amine HM, Othmane A, Saad M. The impacts of control systems on hybrid energy storage systems in remote DC-microgrid system: A comparative study between PI and super twisting sliding mode controllers. Journal of Energy Storage. 2021:103586

[33] Abdelhalim T, Kouider L, Abdelkader R, Hartani MA. Enhanced Vdc bus stability for PV and battery systems through an optimized FOPID controller using a bidirectional DC-DC converter. In: 2023 Second International Conference on Electronics, Energy and Measurement (IC2EM). Medea, Algeria: IEEE; 2023. pp. 1-6

[34] Barakat M. Novel chaos game optimization tuned-fractional-order PID fractional-order PI controller for load–frequency control of interconnected power systems. Protection and Control of Modern Power Systems. 2022;**7**. DOI: 10.1186/s41601-022-00238-x

[35] Abumeteir HA, Vural AM. Design and optimization of fractional order PID controller to enhance energy storage system contribution for damping low-frequency oscillation in power systems integrated with high penetration of renewable sources. Sustainability. 2022; **14**:5095

[36] Ahmed Y, Hoballah A, Hendawi E, Al Otaibi S, Elsayed SK, Elkalashy NI. Fractional order PID controller adaptation for PMSM drive using hybrid grey wolf optimization. International Journal of Power Electronics and Drive Systems. 2021;**12**:745

[37] Haji Haji V, Monje CA. Fractional-order PID control of a chopper-fed DC motor drive using a novel firefly algorithm with dynamic control mechanism. Soft Computing. 2018;**22**: 6135-6146

[38] Zhou Y. A summary of PID control algorithms based on AI-enabled embedded systems. Security and Communication Networks. 2022;**2022**

[39] Nasir M, Saloumi M, Nassif AB. Review of various metaheuristics techniques for tuning parameters of PID/ FOPID controllers. In: ITM Web of Conferences. Rabat, Morocco: EDP Sciences; 2022. p. 01002

[40] Joseph SB, Dada EG, Abidemi A, Oyewola DO, Khammas BM. Metaheuristic algorithms for PID controller parameters tuning: Review, approaches and open problems. Heliyon. 2022

[41] Abushawish A, Hamadeh M, Nassif A. PID controller gains tuning using metaheuristic optimization methods: A survey. International Journal of Computers. 2020;**14**:87-95

[42] Shi J, Song Y. Mathematical analysis of a simplified general type-2 fuzzy PID controller. Mathematical Biosciences and Engineering. 2020;**17**:7994-8036

[43] Wang T, Wang H, Hu H, Lu X, Zhao S. An adaptive fuzzy PID controller for speed control of brushless direct current motor. SN Applied Sciences. 2022;**4**:71

[44] Amine HM, Othmane A, Saad M. The impacts of control systems on

hybrid energy storage systems in remote DC-microgrid system: A comparative study between PI and super twisting sliding mode controllers. Journal of Energy Storage. 2022;**47**:103586

[45] Benbouhenni H, Bizon N, Colak I, Thounthong P, Takorabet N. Application of fractional-order PI controllers and neuro-fuzzy PWM technique to multi-rotor wind turbine systems. Electronics. 2022;**11**. DOI: 10.3390/electronics11091340

[46] Salimi M, Klumpner C, Bozhko S. Sliding mode input current control of the synchronous DC-DC Buck converter for electro-mechanical actuator emulation in more electric aircrafts. Energies. 2022;**15**. DOI: 10.3390/en15249628

[47] Abianeh AJ, Ferdowsi F. Sliding mode control enabled hybrid energy storage system for islanded dc microgrids with pulsing loads. Sustainable Cities and Society. 2021;**73**: 103117

[48] Xiong L, Li P, Wang J. High-order sliding mode control of DFIG under unbalanced grid voltage conditions. International Journal of Electrical Power & Energy Systems. 2020;**117**:105608

[49] Ibrahim B, Abdelkader H, Hartani MA, Kayisli K. Optimization of sliding mode control for doubly fed induction generator systems using particle swarm and Grey wolf algorithms. Electric Power Components and Systems. 2023:1-14

[50] Bekraoui F, Harrouz A, Roummani K, Boussaid I, Bekraoui A. PS and GW optimization of variable sliding gains mode control to stabilize a wind energy conversion system under the real wind in Adrar, Algeria. International Journal of Nonlinear Sciences and Numerical Simulation. 2022;**24**

[51] Mohapatra B, Sahu BK, Pati S. A novel optimally tuned super twisting sliding mode controller for active and reactive power control in grid-interfaced photovoltaic system. IET Energy Systems Integration. 2023;**5**:491-511. DOI: 10.1049/esi2.12117

[52] Debdouche N, Deffaf B, Benbouhenni H, Laid Z, Mosaad MI. Direct power control for three-level multifunctional voltage source inverter of PV systems using a simplified super-twisting algorithm. Energies. 2023;**16**: 4103

[53] Manzanilla A, Ibarra E, Salazar S, Zamora ÁE, Lozano R, Muñoz F. Super-twisting integral sliding mode control for trajectory tracking of an unmanned underwater vehicle. Ocean Engineering. 2021;**234**:109164

[54] Souissi A. Adaptive sliding mode control of a PEM fuel cell system based on the super twisting algorithm. Energy Reports. 2021;**7**:3390-3399

[55] Zhang X, Hu W, Wei C, Xu T. Nonlinear disturbance observer based adaptive super-twisting sliding mode control for generic hypersonic vehicles with coupled multisource disturbances. European Journal of Control. 2021;**57**: 253-262

[56] Gasmi H, Mendaci S, Laifa S, Kantas W, Benbouhenni H. Fractional-order proportional-integral super twisting sliding mode controller for wind energy conversion system equipped with doubly fed induction generator. Journal of Power Electronics. 2022;**22**:1357-1373

[57] Taibi D, Amieur T, Bechouat M, Kahla S, Sedraoui M. State feedback control of DC-DC converter using LQR integral controller and Kalman filter observer. In: International Conference on Digital Technologies and

Applications. Fez, Morocco: Springer; 2021. pp. 1699-1709

[58] Ferahtia S, Djeroui A, Mesbahi T, Houari A, Zeghlache S, Rezk H, et al. Optimal adaptive gain LQR-based energy management strategy for battery–supercapacitor hybrid power system. Energies. 2021;**14**:1660

[59] Neacsu DO, Sirbu A. Energy savings with LQR control of dc/dc converters. In: IECON 2018-44th Annual Conference of the IEEE Industrial Electronics Society. Washington, DC, USA: IEEE; 2018. pp. 1198-1203

[60] Benhammou A, Hartani MA, Tedjini H, Rezk H, Al-Dhaifallah M. Improvement of autonomy, efficiency, and stress of fuel cell hybrid electric vehicle system using robust controller. Sustainability. 2023;**15**:5657

[61] Benhammou A, Tedjini H, Hartani MA, Ghoniem RM, Alahmer A. Accurate and efficient energy management system of fuel cell/battery/supercapacitor/AC and DC generators hybrid electric vehicles. Sustainability. 2023;**15**:10102

[62] Benhammou A, Tedjini H, Guettaf Y, Soumeur MA, Hartani MA, Hafsi O, et al. Exploitation of vehicle's kinetic energy in power management of tow-wheel drive electric vehicles based on ANFIS DTC-SVM comparative study. International Journal of Hydrogen Energy. 2021;**46**: 27758-27769

[63] Benhammou A, Hartani MA, Tedjini H, Guettaf Y, Soumeur MA. Breaking new ground in HEV energy management: Kinetic energy utilization and systematic EMS approaches based on robust drive control. ISA Transactions. 2024;**S0019-0578**(0024): 00047

[64] Subha S, Nagalakshmi S. Design of ANFIS controller for intelligent energy management in smart grid applications. Journal of Ambient Intelligence and Humanized Computing. 2021;**12**: 6117-6127

[65] Aloo LA, Kihato PK, Kamau SI, Orenge RS. Modeling and control of a photovoltaic-wind hybrid microgrid system using GA-ANFIS. Heliyon. 2023;**9**

[66] Hartani MA, Hamouda M, Abdelkhalek O, Mekhilef S. Sustainable energy assessment of multi-type energy storage system in direct-current-microgrids adopting Mamdani with Sugeno fuzzy logic-based energy management strategy. Journal of Energy Storage. 2022;**56**:106037

[67] Vivas FJ, Segura F, Andújar JM, Palacio A, Saenz JL, Isorna F, et al. Multi-objective fuzzy logic-based energy management system for microgrids with battery and hydrogen energy storage system. Electronics. 2020;**9**. DOI: 10.3390/electronics9071074

[68] Abdeselem C, Othmane A, Brahim G, Amine SM, Oussama H, Hartani MA. Power management strategy based sugeno fuzzy logic rules in an electric wheelchair. International Journal of Power Electronics and Drive Systems. 2021;**12**:1187

[69] Abdolrasol MG, Ayob A, Mutlag AH, Ustun TS. Optimal fuzzy logic controller based PSO for photovoltaic system. Energy Reports. 2023;**9**:427-434

[70] Shafei MAR, Ibrahim DK, Bahaa M. Application of PSO tuned fuzzy logic controller for LFC of two-area power system with redox flow battery and PV solar park. Ain Shams Engineering Journal. 2022;**13**:101710

[71] Tran TA. The optimization of marine diesel engine rotational speed control process by fuzzy logic control based on particle swarm optimization algorithm. Future Internet. 2018;**10**:99

[72] Ali MN, Mahmoud K, Lehtonen M, Darwish MM. Promising MPPT methods combining metaheuristic, fuzzy-logic and ANN techniques for grid-connected photovoltaic. Sensors. 2021;**21**:1244

[73] Alhato MM, Ibrahim MN, Rezk H, Bouallègue S. An enhanced DC-link voltage response for wind-driven doubly fed induction generator using adaptive fuzzy extended state observer and sliding mode control. Mathematics. 2021; **9**. DOI: 10.3390/math9090963

[74] Shan Y, Hu J, Li Z, Guerrero JM. A model predictive control for renewable energy based AC microgrids without any PID regulators. IEEE Transactions on Power Electronics. 2018;**33**:9122-9126

[75] Akbari MAH, Farzinfar M, Sadeghi M, Mousavi SS. Optimal model predictive based on super-twisting fractional order sliding mode control to regulate DC-link voltage of DC microgrid. Journal of Engineering Research. 2022. DOI: 10.36909/jer.16267

[76] Nassef AM, Abdelkareem MA, Maghrabie HM, Baroutaji A. Metaheuristic-based algorithms for optimizing fractional-order controllers—A recent, systematic, and comprehensive review. Fractal and Fractional. 2023;**7**. DOI: 10.3390/fractalfract7070553

[77] Alilou M, Azami H, Oshnoei A, Mohammadi-Ivatloo B, Teodorescu R. Fractional-order control techniques for renewable energy and energy-storage-integrated power systems: A review. Fractal and Fractional. 2023;**7**:391

[78] Hartani MA, Messaoud H, Othmane A, Saad M. Contribution to strengthening bus voltage stability and power exchange balance of a decentralized DC-multi-microgrids: Performance assessment of classical, optimal, and nonlinear controllers for hybridized energy storage systems control. Sustainable Cities and Society. 2023;**96**:104647

[79] Ozdemir S, Altin N, Sefa I, Zhang Z, Komurcugil H. Super twisting sliding mode control of three-phase grid-tied neutral point clamped inverters. ISA Transactions. 2022;**125**:547-559

[80] Agrawal N, Samanta S, Ghosh S. Modified lqr technique for fuel-cell-integrated boost converter. IEEE Transactions on Industrial Electronics. 2020;**68**:5887-5896

[81] Alquennah AN, Trabelsi M, Krama A, Vahedi H, Mohamed-Seghir M. ANN based auto-tuned optimized FCS-MPC for grid-connected CSC inverter. In: 2022 3rd International Conference on Smart Grid and Renewable Energy (SGRE). Doha, Qatar: IEEE; 2022. pp. 1-6

[82] Lü X, He S, Xu Y, Zhai X, Qian S, Wu T, et al. Overview of improved dynamic programming algorithm for optimizing energy distribution of hybrid electric vehicles. Electric Power Systems Research. 2024;**232**:110372

[83] Ma Y, Ma Q, Liu Y, Gao J. Adaptive optimization control strategy for electric vehicle battery thermal management system based on pontryagin's minimal principle. IEEE Transactions on Transportation Electrification. 2023;**10**

[84] Shayeghi H, Rahnama A, Bizon N. TFODn-FOPI multi-stage controller design to maintain an islanded microgrid load-frequency balance considering responsive loads support. IET Generation, Transmission &

Distribution. 2023;**17**:3266-3285. DOI: 10.1049/gtd2.12898

[85] Nempu PB, Sabhahit JN. Stochastic algorithms for controller optimization of grid tied hybrid AC/DC microgrid with multiple renewable sources. Advances in Electrical & Computer Engineering. 2019;**19**

[86] Jing W, Lai CH, Wong WS, Wong MDJSET. Assessments, dynamic power allocation of battery-supercapacitor hybrid energy storage for standalone PV microgrid applications. Sustainable Energy Technologies and Assessments. 2017;**22**:55-64

[87] Punna S, Manthati UB. Optimum design and analysis of a dynamic energy management scheme for HESS in renewable power generation applications. SN Applied Sciences. 2020; **2**:1-13

[88] Amine HM, Hamouda M, Abdelkhalek O, Benhamou A, Ali B, Mekhilef S. Robust frequency-decoupling-based power split of battery/supercapacitor hybrid energy storage systems in DC microgrids. In: The 1st International Conference on Physics of Semiconductor Devices, Renewable Energies and Environment. Bechar, Algeria: MDPI, Physical Sciences Forum; 2023. p. 6